Copyright © 2017

All rights reserved

PROLOGUE

A strange coincidence happened in my life a few years ago when I decided to go to a holistic dentist in Spokane, Washington.

I wanted to change my amalgams from cavities I acquired years ago into benign epoxies so the mercury wouldn't leach into my brain. Why? I had become convinced that amalgams were dangerous and that I might lose my thought processes to dementia or Alzheimer's as I got older!

After waking that morning the song, "Good Vibrations" by the Beach Boys kept running through my mind. I didn't recall hearing that song in months if not years. Yet it kept ringing in my head while I showered dressed and drove nearly 70 miles to get to the Dentist's office.

His reception area was very much the same as any other dentists with the exception that it had a bookcase with a lot of books related to holistic health and healing.

One immediately caught my eye and I was stunned! The title was "Vibrations"! As I paged through it the topic related to how "vibrations" affect every moment of our lives. One of the themes suggested that certain families in a community were agitated more than normal and had illnesses that others just a few blocks away never had.

The story postulated that there was an underground "river" flowing a few yards under these homes and, after careful scrutiny with electronic instrumentation that "vibrations" or harmonics caused by the sometimes irradical flow of this river caused the symptoms that these families were experiencing. From that moment on I began to research everything I could find in the Physical Sciences, Religion and Medicine related to "vibrations".

Not being a great scholar but having an insatiable appetite for knowledge I always tried to sort out cause and effect throughout my life.

I decided to take a look at life as I knew it and decided that there were forces everywhere that do, indeed, affect us and even their relationships with each other. I decided to go back to basics.

Basics? From the beginning of time!

It would seem that to try and sort things out from the beginning would be audacious without a PhD in a myriad of subjects starting with Physics, Electronics, Chemistry and Earth Science.

Many of the great minds throughout history discovered much about the Universe with little or no formal education. Fortunately they were able to record their findings on every form of writing and Hieroglyphics.

Sorting through some of these findings as well as the exponential advancement of Science and Technology in my life time, having a career in the advent of semiconductors through maturation of that technology my curiosity got the best of me and I started to write this missive.

This is a work of fits and starts and sometimes weeks without putting any words to this as I pondered what and how to make this presentation. I have no agenda but I have some knowledge about how things work. It's called deductive, inductive, abstract reasoning and research abilities! Not many people have them all I have been given some of these gifts, at least in some measure.

We can create. We can invent. We can fix things. We discover. We can make things happen. We can provoke. We can motivate and we can develop and destroy relationships. Most of all the human condition dictates that we must keep seeking, searching and learning. From sailing around the world on reed rafts, old wooden hulled ships, to traveling across nations on horseback just to see what is there or to seek fortune in the bowels of the earth, to flying more than 25,000 miles per hour, to orbiting the earth and moon in manned vessels, to the new ambition of going to Mars! Man is a wonderful creature with an insatiable appetite to accomplish something meaningful.

I have a wonderful and sometimes preposterous story to tell and I hope that you enjoy the ride you are about to get.

Good "GOD" Vibrations

The Essence of Everything
&
Other Observations

In the beginning, from the very essence of God, a blast came forth so powerful, so profound that at least one thing we call a Universe came into being. We have come to know it the Big Bang.

It has a radius and it travels outwards in all directions at a nominal speed of 500,000 miles per hour! 500,000 miles per hour for at least a billion years and even more!

Within this Universe there are untold and unimaginable Galaxy's with stars and distances between them that are unmeasurable! It is a fact that the earth not only rotates at some one thousand miles per hour at the equator it is moving (along with our solar system) at somewhere near 60,000 miles per hour!

As these all careen through the ether it is amazing that there are not more collisions or more chaos that happens or is observable. Even with our ability to reach out into the past with our newer telescopes we only see less than a fraction of sufficient information to have a clue as to the dynamics of this thing we call the Universe, moving at speeds that are unfathomable.

Because of the so-called Laws of Entropy, the Second Law of Thermodynamics, the Universe and all that is motion will, in something we call time, slow down over the future eons of time. We will never see it happen but it must be so!

So powerful was noise at Creation no one can comprehend the magnitude of that noise, actually it was a voice that said "LET THERE BE" and it was so! We are still left in awe as to the energy behind it.

He was satisfied. That was the beginning of the beginning of what we are even to think about the beginning, which we cannot nor ever be able to comprehend.

The noise was so profound that after tens of billions of years we have been able to invent the devices to try and determine how powerful, old and loud that noise is and where it came from!

Today, scientists still do not know and most deny that it was supernatural. Today, they only try to quantify what God has already created or left for us to explore and identify! They and we tend to think that we "discover", "invent" and "quantify" as if we mere mortals were capable of doing anything but, through experiments, calculations and reasoning, only realize that God put everything in perfect order!

Yet, we believers know with a certainty how, what and where this thing we call reality and the Universe is. It is beyond what we call a "miracle" and it is something that we cannot ever fathom entirely in the here and now. God, our Father, has infinite power and glory that we mere mortals can never fathom in our feeble mortal bodies.

He, with the vibration of His essence can move the Universe, make an earthquake, rise up nations, break the earth, make a nova from a star and chastise a people that He especially created for His own pleasure.

With the great noise and creation of the Universe there is a resonance and frequency in every particle of matter from the smallest part of an atom (quark?) to the largest star and everything in between; mineral, vegetable and animal and everything that constitutes a thing!

The Periodic Table that we all came to know in our Chemistry and Physics classes define all known elements to man. There are most certainly more that we have not discovered. From the lightest of all elements, Helium to the heaviest of all elements they are all constructed of atoms and have valences. Some bond together in such ways as to form molecules and structures. From a speck of dust to the largest mountain or skyscraper these atomic structures are bonded tightly or loosely, amorphous or crystalline, soft or hard. Different elements are melded together to form other materials or structures such that we can build or manufacture some of the most complex items on the planet.

Several elements are used in combination using heat and electronic bombardment to create our wonderful semiconductors (replacing the vacuum tube at the outset) that have become the heart of our IPods, Computers, Cell Phones, Instrumentation, Guidance Systems and Automobile widgets! It is hard to believe that it has only been a scant 70 years since the first single transistor was forged out of germanium and indium at Bell Labs. This transistor has evolved into substrates smaller than your thumbnail which house more than a million transistors, resistors, capacitors and metallic pathways that are the memories for all of your handheld devices. Simply amazing! Yet, all the while the atomic elements making up these devices are all in motion, oscillating at the speed of light and held together by their valances and electromagnetic energy!

It is hard to comprehend the fact that everything you see and some things that you cannot see are in motion, that they are all oscillating, that most are harmonics emitting an electromagnetic force that is measurable and readable with sophisticated instrumentation! This would include the smallest particle to the largest star and everything in between.

Further, it is interesting to note that amorphous matter in the Universe gravitates to circles and spheres. Circles and spheres? The molten masses that came out of the Creation or Big Bang were, for all intent and purposes, "gooey" before they cooled into stars or their satellites be it planets or moons. All the stars, planets and their satellites (Moons) are round or at least spheres. If they do not or did not rotate they formed nearly perfect circles. If they do or did rotate or acted upon by a nearby body of mass they are likely spheres. As an example soap bubbles are round. We know that water drops are round in space but only elongate while falling as rain. Boiling water bubbles are round and the list is endless. Each element of an atomic structure is round (proton, neutron and electrons). These are immutable laws that God has put into place.

All of mankind's endeavors to create and discover are only to quantify what God has created to begin with and put into motion. We are only able to mix and match that which was given to us to bring about the materials we use to build and concoct. At the same time God gave us the intellect to use his creations in useful and meaningful ways to make our lives easier, safer and more comfortable. And, of course, we have used these elements to create and manufacture the most dangerous weapons mankind has ever seen, from hardware, chemical and biological to atomic which can annihilate tens of thousands in an instant of time! At the same time atomic energy can provide decades of energy we use to heat and cool our homes, propel our ships at sea and run entire cities with little safety hazard or pollution.

There are no straight lines in the Universe, on the planet or in the natural order of things. We try to organize and make things straight, right, logical and meaningful. God does no such thing. The Universe is organized chaos but perfectly organized.

A contradiction in terms? Not at all!

Herein are facts and ideas that defy the imagination and you will forever rethink how and why things work in this thing we call time and space. You will find redundancy throughout this missive. These are purposely repeated to reinforce the information into your psyche and it is done with all good intentions.

SIMPLE ATOM EXAMPLE

Quantum physicists discovered that physical atoms are made up of vortices of energy that are constantly spinning and vibrating, each one radiating its own unique energy signature. Therefore, if we really want to observe ourselves and find out what we are, we are really beings of energy and vibration, radiating our own unique energy signature - this is fact and is what quantum physics has shown us time and time again. We are much more than what we perceive ourselves to be. If you observed the composition of an atom with a microscope you would see a small, invisible tornado-like vortex, with a number of infinitely small energy vortices called quarks and photons. These are what make up the structure of the atom. As you focused in closer and closer on the structure of the atom, you wouldn't see anything, you would observe a physical void. The atom has no physical structure, we have no physical structure and physical things really don't have any physical structure! Atoms are made out of invisible energy, not tangible matter. Even though our linear accelerators can break loose an electron from an atom and move it to near speeds of light to make an imprint on a material that is measurable it seems to have a physical structure. In reality it is pure energy or electromotive force.

It's quite the conundrum, isn't it? Our experience tells us that our reality is made up of physical material things, and that our world is an independently existing objective one. The revelation that the universe is not an assembly of physical parts, suggested by Newtonian physics, and instead comes from a holistic entanglement of immaterial energy waves stems from the work of Albert Einstein, Max Planck and Werner Heisenberg, among others

"If you want to know the secrets of the universe, think in terms of energy, frequency and vibration." - Nikola Tesla

Operating frequencies from Extremely Low (ELF) to extremely high (Gamma) is nearly all that modern instrumentation can identify. There definitely are higher and lower frequencies that have not been identified or discovered. These frequencies are oscillations or "vibrations" and "harmonics" that interact with one another. At the atomic level every structure, particle and molecule has similar but different frequencies that make them independent, sometimes cohesive and literally make up all matter.

The fact is we are all interdependent on the oscillations (vibrations) emanating from every particle of matter. Our brains emit electrical impulses as well as every other part of our body. *All physical reality is made up of vibrations of energy. Your thoughts too are vibrations of energy. This is not a concept or theory, but rather the startling new reality that quantum physics now reveals to us.* Taking this fact to a larger level the earth, sun, moon and every other star emits larger electromagnetic signatures, all measurable. Without these interacting with each other we mere humans, indeed all life forms, would not be able to survive

Experiments were done a few years back whereby rabbits were placed in a Faraday cage with all the water and food they could ingest over weeks of time. Because the Faraday cage shields all electromagnetic and electromotive forces the rabbits died within a few days proving that they were dependent on the "vibrations" or oscillations from all matter required for life. This experiment was performed several times using different breeds of rabbits with the same results.

We have all come into contact with people that we seem to have an affinity. No, I am not necessarily referring to the attraction the opposite sexes have for one another but a different thing. Conversely we have all come into contact with people that seem to agitate us or make us feel uncomfortable. Sometimes their negative energy is unsettling to us and we can't get away from this person or thing fast enough. This is easily explained with the interactions of each of our emissions or electromotive emissions that we all have. Some are more compatible than others with each person we meet. Naturally, most encounters with others are neutral with no sense of positive or negative feeling.

Many studies have been done with the emission of Aura's from various life forms including individuals. Those who are extremely sensitive to the Aura's of others claiming that they are positive, negative, large or small. Instrumentation exists which can actually measure these "Aura's". What Aura's actually are is the emission of electromagnetic/electromotive force (vibrations) from an individual.

Phantom Limb Pain (PLP) is another phenomenon that can be attributed to vibrations/oscillations/electromotive force. The adult amputee still senses the limb either by pain or itching. The cellular interaction within the body attempts to adjust to the lack of emission from the absent limb. Over time, however, an adjustment is made and these PLP's cease.

All cells first communicate by light and electrical impulse which triggers chemical activity releasing hormones, adrenaline and a host of other body activities which sustain a healthy life. Again, all are "vibrations", oscillations and electro activities.

Earth's Electromagnetic Fields & How They Connect To Our Own

Science has recently shed light on the fact that what we used to perceive as 'human' aura is actually real. All of our bodies emit an electromagnetic field, and this fact plays a very important role far beyond what is commonly known when it comes to understanding our biology, and the interconnectedness we share with all life.

For example, did you know that **the heart emits the largest electromagnetic field of all the body's major organs**? These fields and the information encoded into them can change based on how we are feeling, what we thinking, and different emotions we take on. The heart even sends signals to the brain through a system of neutrons that have both short-term and long-term memory, and these signals can affect our emotional experiences. The emotional information that's modulated and coded into these fields changes their nature, and these fields can impact those around us. As Rolin McCratey, Ph.D, and director of research at The HeartMath institute tell us, "we are fundamentally and deeply connected with each other and the planet itself."

Research findings have shown that as we practice heart coherence and radiate love and compassion, our heart generates a coherent electromagnetic wave into the local field environment that facilitates social coherence, whether in the home, workplace, classroom or sitting around a table. As more individuals radiate heart coherence, it builds an energetic field that makes it easier for others to connect with their heart. So, theoretically it is possible that enough people building individual and social coherence could actually contribute to an unfolding global coherence.

The quote above comes from Dr. Deborah Rozman, the President of Quantum Intech. We are living in exciting times when it comes to science, and although not emphasized and studied in the mainstream as much as we'd like, science is acknowledging that we are all part of a giant web of connections that, not only encompasses life on this planet, but our entire solar system and what lies beyond it.

So, what exactly is heart coherence? Well, it implies order, structure, and as Dr. Rozman puts it, "an alignment within and amongst systems – whether quantum particle, organisms, human beings, social groups, planets or galaxies. This harmonious order signifies a coherent system whose optimal functioning is directly related to the ease and flow in its processes." Basically, feelings of love, gratitude, appreciation and other 'positive' emotions not only have an effect on our nervous system, but they have an effect on those around us, far beyond what we might have previously thought.

It is astounding that our heart beats thousands of times a day and pumps gallons of our blood in a perfect cycle through arteries, veins and capillaries receiving oxygen and releasing carbon dioxide! The energy to accomplish this is simply amazing! Where does this energy come from? Where does all the energy that comprises the universe come from? Is it self perpetuating? Emphatically no! This energy is the essence of God and He directs every heart heartbeat from His throne room!

Another point that illustrates the importance of coherence is the fact that several organizations around the world have conducted synchronized meditations, prayers, intention experiments, and more. A number of studies have shown that collective meditations, prayer or focused intention directed toward a certain positive outcome can have measurable effects.

For example, one study was done during the Israel-Lebanon war in the 1980s. Two Harvard University professors organized groups of experienced meditators in Jerusalem, Yugoslavia, and the United States with the specific purpose of focusing attention on the area of conflict at various intervals over a 27-month period. During the course of the study, the levels of violence in Lebanon decreased between 40 and 80 percent each time a meditating group was in place. The average number of people killed during the war each day dropped from 12 to three, and war-related injuries fell by 70 percent.

Every individual's energy affects the collective field environment.

- The Earth's magnetic fields are a carrier of biologically relevant information that connects all living systems

- Earth has several sources of magnetic fields that affect us all. Two of them are the geomagnetic field that emanates from the core of the Earth, and the fields that exist between Earth and the ionosphere. These fields surround the entire planet and act as protective shields blocking out the harmful effects of solar radiation, cosmic rays, sand, and other forms of space weather. Without these fields, ice as we know it could not exist on Earth. They are part of the dynamic ecosystem of our planet

  These energetic fields are known to scientists, but there are still many unknowns. Solar activity and the rhythms taking place on Earth's magnetic fields have an impact on health and behavior. This is firmly established in scientific literature.

Scientific literature also firmly establishes that several physiological rhythms and global collective behaviors are not only synchronized with solar and geomagnetic activity, but that disruptions in these fields can create adverse effects on human health and behavior.

When the Earth's magnetic field environment is distributed it can cause sleep problems, mental confusion, usual lack of energy or a feeling of being on edge or overwhelmed for no apparent reason. At other times, when the Earth's fields are stable and certain measures of solar activity are increased, people report increased positive feelings and more creativity and inspiration. This is likely due to a coupling between the human brain, cardiovascular and nervous system with resonating geomagnetic frequencies.

The Earth and ionosphere generate frequencies that range from 0.01 hertz to 300 hertz, some of which are in the exact same frequency range as the one happening in our brain, cardiovascular system, and autonomic nervous system. This fact is one way to explain how fluctuations in the Earth's and Sun's magnetic fields can influence us. Changes in these fields have also been shown to affect our brain waves, heart rhythms, memory, athletics performance, and overall health. Changes in the Earth's fields from extreme solar activity have been linked to some of humanity's greatest creations of art, as well as some of its most tragic events.

Research is indicating that human emotions and consciousness ecocide information into the geomagnetic field and this encoded information is distributed globally. The Earth's magnetic fields act as carrier waves for this information which influences all living systems and the collective consciousness.

"God didn't just drop us off on earth!" He left us with all the natural resources of plants and animals for our nourishment and an unending supply of minerals; oil, coal, copper, iron and trees on every livable continent for us to exploit for our convenience and use!

Because of the perfect designs and synergism that exists in our world evolution is impossible. No matter what Scientific American, Discover, National Geographic or any other so-called Academic expert may postulate or claim as fact! There is absolutely no proof that we evolved from some sort of primordial ooze to a multi-cellular creature, emerging from the oceans, crawling first, then standing upright and wandering for eons of time until our brain began to intelligently function!

Our human physical and physiological attributes prove beyond any shadow of a doubt that we are a creation of near perfection and not a random happenstance of evolution. We have ears on either side of our head to hear from all directions. We have two frontal eyes so that we have near 180 degree vision (including periphery) with which to determine distance and focus, near and far. We have limbs placed perfectly to perform ambulation, dexterity and reach for nearly everything activity we require.

Our organs and body ecology are in harmony with glans and other secretions which metabolize our food, adrenaline to achieve near super human activity and a plethora of other subtleties. We can reason, dream, think in the abstract, imagine and create. We can love, respect our dead, fight, flee and have compassion. No other creatures on the face of the planet can do any of these things! We were created here with all of the necessary accoutrements to live a healthy and robust life.

Nicolai Tesla discovered the electromotive/vibration/oscillating forces emanating in the Universe and applied them in his state-of-the-art experiments. His inventions and technological advances under the financial umbrella of George Westinghouse exceeded those of Thomas Edison and is the only reason the United States is on Alternating Current Power instead of Direct Current Power as proposed by Edison.

Moreover Tesla was able to develop the equipment that allowed for near free energy which has been suppressed since then by those who enjoyed immense profits from the sources of energy which are prevalent today!

It's an electronic universe. This is far more than the cold, gravity-driven machine. Electromagnetic forces are 1,000 billion billion billion billion times stronger than gravity (10^{39}) and much more realistic as the force which actually holds our universe together. Space is filled with electromagnetic energies, clouds have been photographed by The Hubble and, of course, radio telescopes.

Electricity and magnetism are out there! Whereas gravity has never been seen, felt or photographed. Catalogues of the General Electric Company in the late 19th century had a number of electronic healing devices. Its 1893 catalogue illustrates nine magnetos, ranging from one 20 centimeters long, boxed in pine, to the Phoenix, boxed in mahogany with a dial "to measure strength". Assorted electrodes "for foot, tooth, and ear, with plated handles" were available for the device.

Another type of device was the induction coil, originally developed to detonate explosive charges. By 1888, GEC was offering induction coil apparatus for medical use, complete with bichromate battery, in a wooden case. By 1890 their range had grown to 10 models, with many variants, and electrodes engineered to treat particular parts of the body, from eye muscles to the spine.

The formation of the Food and Drug Administration and the American Medical Association in 1933 for the so-called purposes of protecting the public from spurious medical claims and treatments resulted in the "protection racket" for their trained doctors and own Pharmaceuticals. They instantly attacked viable electronic and natural cures for diseases that were well proven over time and, in the end garnered arrest and fines power delved out to all who didn't capitulate to their demands. Today, nearly 90 years later they have a stronger hold and more power than ever before against alternative medicine and procedures, even if it works!

Angels, Demons, Ghosts, Poltergeists and Pathogens

Between the movies, Coast to Coast late night on the radio and television programs about the paranormal we are left with lingering notions about ghosts, spirits and even Angels and Demons. In fact, a movie based on Dan Browns book by that name was made starring Tom Hanks.

Do supernatural entities exist? Why are they not around us all the time, at least where we can all "feel" them or sense them or even see them? After all, in the ancient of days there seems to have been many cultures and civilizations that lent credence to these "spirits" and many nations had them as gods.

If they are around from time to time, why is it that only a very few people have contact with them and then not all the time. And why is it that the most sophisticated equipment in the world cannot sense, see or alert us as to the presence of spirits, if they indeed exist at all.

From the Bible we know that Satan, Gabriel, Michael as well as many other so-called messengers have transcended the eternal continuum to the time space continuum many times. We read it in Genesis, Job, Isaiah, Daniel, the Gospels and Revelation. These are all arch-angels or simply angels. The Catholic Bible gives us a glimpse of a couple others. Some are called simply messengers which many preachers claim that they weren't angels at all. And, of course, Egyptian, Roman, Greek and Nordic mythology tells us of the Gods that sometimes visit and perform various feats and acts in the heavenlies.

The Bible tells us that, as the result of the fall of Satan, or Lucifer, that he took 1/3 of the angels with him to do his bidding. That would be to thwart the efforts of Yahweh God on this planet through His peculiar and particular people, we White, Israelites of the earth. One of the first things that we know is that Satan did fornicate with Eve in the garden bringing about Cain or the Kennite line of peoples. His fraternal twin was Abel which was the offspring of Adam and Eve and was born from a separate zygote.

Secondly, the Bible tells us that the angels left their first state (eternal) and came into the time/space continuum and fornicated with the women on earth whom they found beautiful. These resulted in the so-called Nephilim or Zamzummim or giants of the earth. Whether this means men and women of great power as some claim or great stature as literal giants, we know that there were giants in the Bible and their bones have been found around the world…………some over 15 feet tall! Most notable in the Bible are Og and Goliath who were still living at the time of King David, about 1000 BC.

Remember when the angel of the lord was to meet with Daniel and had to contend with the Prince of Persia for 21 days until Michael the arch-angel intervened for the "messenger" that came to Daniel. After speaking to Daniel this "entity", which Daniel called in "human form", not entirely human, said he had to go back to continue his fight against the Prince of Persia, a fascinating story.

Based upon these Biblical accounts we know that these spiritual entities somehow make their way from the spirit world into ours, at least they used to. Where are they, then, in the spirit world, up there above us beyond the moon, in the center of the earth or the netherworld beneath us? It is my belief that they are all around us.

Let's start with the basics and bring you along in a logical, rational way so that you are convinced that these do exist and are all around you all of the time. Scary? No, not really because these entities rarely transcend the eternal continuum to the time space continuum. Why? Well, we cannot know at this time. We just know that they don't.

The foundation of all matter has to deal with atoms, that is to say protons neutrons and electrons and even smaller particles. These all "vibrate" or oscillate if you will. In fact, all matter has a harmonic. From the smallest particle to the largest star including everything in between; molecules, tissues, organs, eyes, the body of each and every creature, rocks, water, air and the earth has an electronic signature or oscillating frequencies or harmonic and most are measurable or identifiable by their frequencies.

We know that Gods Laws are immutable and they do not or cannot change unless acted upon by outside forces. If this were not so everything in the universe and around us would be random and chaotic. We couldn't exist in the state that we do nor could anything else. That is why evolution is impossible.

A brief note about Crystals is appropriate. Crystals are used to emanate radio waves from the radio stations all over the country. These will generate the carrier wave that houses the signal that we hear on our radios. For example, you can hear this when your radio station has so-called dead air, when there is a long gap between music playing or someone talking and you hear a sort of static. That static is either the carrier wave carrying nothing or, sometimes just electro-magnetic interference,

If you could squeeze a crystal sufficiently or place a crystal under extreme pressure without shattering it, the crystal will generate power that can be measured. Such is the essence of a crystal. In the so-called occult or Metaphysical world many adherents wear crystals around their neck and place them in strategic locations around their houses. They claim healing powers and even say the crystals enhance the ability to communicate with the spirit world. True or not there is a nice business in mining various types of minerals for crystals, shaping them and selling them to those who believe. The church of today claims this practice to be demonic and worse and forbid their flocks from wearing crystals for that purpose. It would then be forbidden to wear diamonds, a very special crystal!

So what is the point of all of this? I am going to summarize a lecture I gave in Tucson in 2012. My lecture had to do with Biological Medicine, a discipline for which I am certified. I studied in Canada, worked under the tutelage of a 20 year practitioner and purchased equipment manufactured in Germany that is sold throughout Russia and many other European countries who use it to literally cure cancers of all types and many other maladies that shorten the life of us humans.

This is accomplished by killing or neutralizing the causal agents of these diseases, pathogens. As it turns out all pathogens have an oscillating frequency between 80 and 800 kilohertz. By targeting the exact harmonic or frequency of a particular pathogen, inverting the wave and ever so slightly increasing the amplitude and returning it to the patient, the pathogen is destroyed. This can be accomplished within any human being or other living thing for that matter. It is only a question of identifying that particular pathogen. This can be accomplished with some rather sophisticated electronics equipment, comparators and easily purchased in Germany. Sometimes a simple treatment can be performed with a common 9 volt battery.

All viruses, most bacteria, molds and fungus are harmful to us humans. By killing these we can allow the body's own immune system to bring back a healthy individual, in perfect harmony. That is the essence of all health, the body being in harmony internally. I know it sounds somewhat farfetched. If this was so and it did actually work (and it certainly does) why don't physicians use these procedures to help their patients?

First, they can only do what they are taught in medical school and what is approved by the AMA and FDA. There is no profit in health, only in disease. Using those procedures would eliminate 75% of all pharmaceutical products on the market and 2/3 of all doctors unless they are highly specialized.

Of course, the FDA and the AMA call what alternative practitioners do quackery because they want you to use their spurious methods and drugs in an attempt to affect a cure, procedures that are far less successful than what Alternative practitioners use. They attack the cause, not the symptom.

All of the senses that we have come to know; touch, smell, sight, hearing, taste and the rest are a result of the reception of the vibrations or oscillating frequencies that emit or receive those phenomena.

Our eyes see nothing. The light spectrums are received in the retina of the eye as oscillations. These oscillations are transmitted to the brain and the brain then creates the image which we come to know as sight. Every creature on the planet has been given certain senses that we do not have, receptors that sense on higher or lower frequencies or harmonics than we have.

As an example a cat can see better at night than we can. A dog can smell things that we can't smell and has more acute hearing than we do. An eagle can see far further than we can and can even see fish under the water at these same distances. We all know that an owl sees better at night than we can.

How do you suppose that these creatures, and I've only named a few, are able to do these things? Simply, they are engineered by God to have certain senses on a different level than we humans. Their systems operate having different receptors than we have. They are able to sense levels of oscillation or frequencies higher or lower than we are able.

Moving from long wave lengths or ELF (extremely low frequency or radio waves) towards gamma waves you will see that the frequency or wave length is compressed significantly. You will notice that all measurements cease at the gamma wave length or 10 to the -12 in frequencies. These frequencies are all measurable or extrapolations. The visable light spectrum for we humans is very narrow but large enough for other creatures to see what we cannot. Our senses are affected by oscillations at either end of the light spectrum. In concentration some of these oscillations are capable of killing life or rendering living organisms lifeless!

What if you were to double or even triple the frequency amplitude, shorten the wave length significantly? Think about it. Let's go even further. What if we were to magnify the frequency by a hundred or a thousand times more than the Gamma spectrum! Those possibilities are endless. If possible would that account for parallel universes or..................the spirit world?

Remember, the Bible tells us that God says that one day is as a thousand years. Using simple math, by extrapolation one day of our lives is equivalent to 4 seconds in the eternal continuum if this is so. That would mean that the spirit world can move about all around us at speeds that we cannot fathom.

And, if we were to conclude that matter in that eternal continuum is far larger or smaller it is even possible to conclude that the so-called spirits or angels or demons or ghosts or whatever can even be passing through us each and every nanosecond. That's a billionth of a second!

Jesus confirmed the existence of ghosts when he miraculously appeared to the disciples, including Thomas (who later became known as "Doubting Thomas), entering the room they were in through a wall. There were two significant things that happened here. One, that Jesus confirmed the possibility of ghosts by declaring that the disciples looked as though they had seen a ghost and two, that Jesus seemingly entered the room through the wall, transcending the immutable laws of physics, that is, unless he broke the wall down (Matter cannot occupy the same space at the same time!). Remember, this is after Jesus had been crucified, dead, buried and resurrected. He, obviously, was able to transcend from the eternal continuum at will and defy the "immutable laws of physics."

So there are angels and ghosts and even demons. They function at a different oscillation or frequency than we do, so much more that we cannot see hear or feel. Occasionally, a dog or cat may get a sense of them because of their abilities to hear and see beyond what we are able, probably when these entities are operating or attempting to transcend from the eternal. Proof? I have none and no one else really does either.

These are speculations using the immutable laws of the Universe, or God's laws of the universe to extrapolate how these things exist and where they might exist. Based upon this you can draw your own conclusions about the spirit world.

Always remember, there is no death in the spirit world and, the truth be known, we do not die either. We leave our earthly bodies and translate into the spirit or eternal continuum. Nowhere in the Bible is there any indication there is death or annihilation in the spirit world, including us after we leave our earthly bodies. God and Jesus places spirits and angels in the Abyss, chains them, leaves them in paradise, sends them (or us) to the fire, the outer darkness or allows us to be with him throughout the Universe or in the throne room or living in a room in the mansion that He has for us.

There is a lot that we know and much more that we don't know. We may be assured, however, that one day on that great day, we will all know.

How Biophotons Show That We Are Made Of Light

A biophoton or Ultra-weak Photon Emission, (UPE) is a kind of light particle that is emitted by all living things. Though it exists in the visible and ultraviolet spectrum, in order for us to see it, our eyes would have to be about 1,000 times more sensitive. While we can't see them with our eyes alone, technology has given us a glimpse and what it's shown us may have a profound impact on us all.

Biophotons were first thought to be merely the byproduct of metabolic chemical reactions. That idea is being challenged with an exciting theory that claims biophotons have a much larger role to play when it comes to our physiology and quite possibly our consciousness as well.

Experiments are showing that biophotons (UPEs) can be captured and stored inside of cells and can even travel through our nervous system; suggesting that biophotons might provide a way for cells to transfer energy and communicate information. It's has also been suggested that UPEs might even have properties which help us to visualize images. This makes sense considering how we're creating computers. All computers are is silicon crystal chips which we pass light through to relay 0's and 1's. (Light on, Light Off). The first computer was literally built with a light bulb and these punch-cards which had holes in them, which is how we would program computers.

Today, our computers are far more advanced, and yet at the core, the electrical information we are passing through computers today is still a form of light. Now our sciences are revealing humans work the same way, and Light carries information through our brain, nervous system, and even our DNA.

Scientists are finding that our DNA is a strong source of UPEs, it communicates with and is created from light itself! It's been observed that

DNA produces extremely high biophoton emissions and has excimer laser-like properties. Excimer lasers (or exciplex lasers) are special lasers consisting pseudo-molecules that only exist in a highly excited state and emit light in the ultraviolet range.

Scientists have also discovered that not only do we emit light, we have the ability to effect it with our thoughts alone. In a recent study, participants were placed in a darkened room and asked to visualize a bright light. When they did this, they were able to increase their levels of biophoton emissions significantly, showing that our intentions have an influence on light itself!

Light appears to be a fundamental part of our being. It's hard-coded into our very bodies to function directly with, and through – light.

Authors Comment:

The most powerful force in the Universe is a photon or light. This has been easily proven by driving light into a crystal, exciting that light with high energy electronics and releasing it in a burst. This light can destroy a tank, airplane and precisely etch metal or any other element including diamonds! They have called this tool a LASER!

The following was released to the Internet on February 25, 2016:

TRAVEL TO MARS IN THREE DAYS!

The key to making this happen is photon propulsion, which would use a powerful laser to accelerate spacecraft to relativistic speeds, said Philip Lubin, a physics professor at the University of California, Santa Barbara. "There are recent advances which take this from science fiction to science reality," Lubin said at the 2015 NASA Innovative Advanced Concepts (NIAC) fall symposium last October. "There's no known reason why we cannot do this. - See more at:
http://www.space.com/32026-photon-propulsion-mars-three-days.html#sthash.fUAsTXQw.dpu

BIOLOGICAL MEDICINE

German Biological Medicine, in essence, uses the human bodies own information to determine its harmonic balance, ecology, pathogens and then effecting a specific therapy to remedy discrepancies.

The Universe consists of matter. All matter is comprised of atoms and atomic structures. All atoms have protons, neutrons, electrons and many other smaller particles that have been discovered in the last 50 years. Each atom and atomic structure has a specific and particular resonating frequency and harmonic. These can be identified electronically by instruments such as a Frequency Comparator, a Frequency Counter.

From a particle of dust to the most complex organic structures known such as the human body this resonance and harmonic determine its vitality or energy. Further, every component of the body has a specific identifying harmonic that can be measured. For example, the human body structure has a resonance in the 30 - 40 Hertz range. Virus, bacteria, molds and parasites typically resonate from 80 - 800 Kilohertz (cycles per second).

The HIV virus resonates at 365 KHz according to studies completed by Dr. Hulda Clarks while under contract to the United States Government to identify the resonating characteristics of various pathogens. More than 3000 identifying resonances of various pathogens have been mapped.

When the vitality or energy of any part of the body is disrupted the output changes from the "norm". Health and wellbeing can only be achieved when the ecology or terrain of the body is in resonating and harmonic balance. Body Ph (acidity or alkalinity), oxygenation and energy dynamics that are askew or in contradistinction always results in illness, disease or worse. A body in balance and harmony cannot become diseased even if in contact with highly contagious disease or viral outbreaks.

In the early 1920's Royal Raymond Rife, a scientist residing in San Diego, California discovered that viruses can be destroyed electronically. He was the first (and last) to invent a light refracting microscope that could be used to actually see live viruses on a microscope slide. It had a 17,000 power whereas today's best light refracting microscopes have no more than 4,000. Today, only electron microscopes can see objects that small and smaller but kills any life form organisms in doing so. Rife identified more than 100 viruses by frequency and color.

He determined that these viruses can be "killed"* by inverting the frequency emitted by the virus, returning this modified frequency back to the virus and was able to watch and film these literally exploding!

This procedure works with bacteria, molds and parasites equally. Rife was heralded as the man who discovered a non-invasive cure for cancer, having proved it with over 300 cancer ridden patients over a period of two years. Prominent physicians from Mayo Clinic and Johns Hopkins recognized Rife's work and he was given accolades, much press notoriety and many awards. Later, the newly formed FDA discredited his work, called it quackery (as they do with all alternative procedures that effect cures) and Rife was forbidden to do any other work.

One would logically conclude that by applying any voltage to a pathogen would destroy it. Unfortunately, the specific voltage to a particular pathogen is required so that other cellular structures are not compromised. As it turns out more is not better.

It has only been in the last 35 years that Rife's work has been revisited by German scientists and doctors working in concert to solve a myriad of health problems that we humans are subject to. Dr. Franz Morrel and Eric Rauch of Germany designed and developed the Mora Supra Bipolar system that is used to determine that vitality of various organs within the body. Various modalities and electronic signatures of pathogens have been programmed in the Mora Super. This work is an extension of prior work done Voll, Vega, Bruggerman, Vincent and others. The Mora of today is the 4th generation of systems that required years of knowledge and practice to achieve results that we are able to do rather simply today.

The work encompassed identifying pathogenic agents, establishing a therapy to treat identified disorders, balancing the energy flows of and between vital organs, neutralizing allergens and addressing certain painful afflictions. Today there are 6,000 specific identifiers and remedies incorporated in the Mora allowing the Practitioner to correct an unbalanced, afflicted body into a more healthy ecology or terrain.

These procedures then allow the immune system to cure the ailment. Remember, no one can cure another. Not a physician, not a Practitioner, not a Shaman, not a psychic, not a "healer", not a minister. The forces of nature, set in place by our Creator and the life force that exists in each of us is the true healer. Our own immune system is the mechanism that effects healing every time. Our work only eliminates the debris and forces that act on the immune system, overwhelming it's function, thus preventing it from performing it's function.

Biological medicine is a quantum leap forward in human organ testing and therapy. When resonating frequencies of a pathogen are detected by the Practitioner using the Mora, the frequency is filtered through the Mora, inverted and returned to the identified pathogen(s), thus destroying it. The result is similar to a mini "explosion" of the pathogen.

The remains are discharged from the body through the typical channels that rid the body of toxins and other debris: the sweat glands, sinus drainage, urine, feces and tears. While undergoing these procedures which sometimes require up to three treatments a day up to 21 days, depending on the severity of the illness, it is imperative that the patient drink at least three liters of pure, non-fluorinated water each day. This will facilitate the discharge of whatever debris caused by the treatments.

The patient feels absolutely nothing during all test and therapy procedures. First, they are non-invasive. Secondly, the voltages applied are so low as to be negligible to the patient but are extremely effective in identifying and destroying the pathogen.

Electronic Acupuncture according to Voll (EAV) is a non-invasive method of reading the body meridians that are the pathways to all organs and other structures in the body. Chinese acupuncture is an art/science form of invasive (needles penetrating the meridians) treatment after a symptom is diagnosed. The acupuncturist can only know the effectiveness of treatment by trial and error by inserting several needles and waiting for the patient to exhibit some sign of relief. Chinese acupuncture is an "open loop system" in that the results of treatment cannot be monitored in real time nor can the practitioner determine if any activity is happening at all.

EAV allows the practitioner to determine the health or vitality of an organ or tissue. He then can establish exactly which pathogen or contra indicator is the causal agent and specifically attack this agent. He can then re-test to determine if these agents still exist after therapy. If any remain he can continue therapy until there are none left, sometimes taking several days of they are very numerous and aggressive in reproduction.

Contrary to tradition or allopathic medicine where the physician "forces" a pharmaceutical or procedure upon the patient to relieve systems or extract problems surgically, electronic medicine works with the body's own information to resolve abnormalities in an organ or tissue. Traditional medicine does not have the capability to ascertain problems manifesting within the body in the very early stages of development. Biological medicine solves this failing and often years in advance of any manifestation of disease. It is at this juncture where Biological Medicine proves itself to be the miracle that it is by thwarting this potential disease from further growth.

Biological Medicine also incorporates light therapies which are the most powerful form of therapy known. The various color spectrums revitalize red blood cells, the immune systems, brain function, wellbeing and moods. These colors are used in the Mora programs to aid in the therapeutic effects accelerating the healing process exponentially.

Marlene Geiser-Wyle has been a Practitioner for more than 25 years and author of the best seller "YOU CAN SURVIVE IT ALL." The book provides naturopathic, dietary and homeopathic remedies for virtually every illness and other malady's that are prevalent in our society. She is credited for effecting cures for more than 2500 patients during her practice including more than 1100 cancer ridden individuals!

It is still conjecture whether viruses actually "live" or are inanimate only affecting the proteins in live cells. Rife observed "live" viruses with his microscope. Viruses require a host to "live" whereas bacteria are alive and live without need of a host.

More on the Mora Therapy

For 150 years we have lived with a world view which is shaped in a significant manner by the knowledge of mechanics and chemistry, the scientific bases that quite substantially determine modern civilization.

In spite of the enormous advances in knowledge in the area of electronics, this usual world view is hardly shaken up. Especially in medicine there are many innovations in diagnostics that show the influence of modern physics. However the new findings from biology about the physical-energetic processes of living systems have hardly found an input into the therapy of diseases.

Whereas classical medicine still limits itself principally to the biochemical basic model of living systems (with some additional knowledge from electro-physiology) that's why disease treatment is essentially achieved only from pharmaco-therapy, or rather with reference to the mechanical perception of surgical therapy strategies. It appears that obvious therapy forms in the area of Regulation Therapies like homeopathy, acupuncture, electro-acupuncture, and finally BioResonance Therapy can also influence the body very effectively toward health and often with much fewer side effects.

MORA BioResonance Therapy represents a modern advancement in the area of acupuncture, electro-acupuncture and homeopathy.

The MORA BioResonance Concept - Scientific Bases

In order to correctly understand and assess the MORA BioResonance Method, which represents a completely new therapy beginning within the framework of biophysical therapies, it is necessary to understand the following basic conditions from modern biophysics:

Ø All biochemical processes in living systems are a result of the *electromagnetic interaction* of the involved reaction partners and the *electromagnetic environment*. All biochemical materials generate molecule specific electromagnetic oscillations, which have a feedback effect with higher coherent *electromagnetic fields* steering the biochemical processes.

Ø *Bio-electromagnetic phenomena* like for example biophotons which originate within the cells (cell nucleus) are verifiable in the living organism – with coherent and mono-chromatic characteristics like a laser – and are sent out in specific fluctuation sequences. Through resonance these biophotons cause the transmission of homogenous photons with correspondingly rectified influences into the adjoining cells, more specifically the biochemical processes in the environment. This statement corresponds to the research of the biophysicist *Fritz-Albert Popp*.

Ø *All living systems are "open" systems. To maintain the energy necessitated by life functions, they need an ongoing supply of signal energy (not chaotic or heat energy), and the influence from physical energy forms, especially from electromagnetic oscillations, which play a very special role in the formation and regulation of the life processes.* According to the statements of the physicist *H. Fröhlich* and physical chemist *I. Prigogine* (Nobel Prize in 1977) *life is only possible through continuous signal energy supply*, which increases the sensitivity of the living systems to the physical stimuli from outside and their controllability with the fixed size of the system. The biological system "person" (very high fixed size) already responds especially sensitively to small sized electromagnetic stimuli which have an effect from the outside and the inside.

Ø The skin with its dependent structures is a multi-layered sense organ and in this position, in the area of specific skin zones (acupuncture points, neural zones), is therefore able to take up various information, even *electromagnetic information*, from the outside world and to process it and pass it on inwards. These zones contain special morphological structures like nerve endings, receptors, etc. and they sensorily connect the body surface with the interior of the body.

Ø All biochemical materials deliver molecule specific electromagnetic oscillations (fields) that can be transferred with or without a cable by radio [*waves*], and influence the living organism according to their characteristics. This knowledge exists in the medication test according to *Dr. Reinhold Voll* and the further investigations of *Dr. Franz Morell* and *Mr. Erich Rasche*.

In this sense the measurement values of the acupuncture points can probably be assessed. Compared with their environment they show a characteristic skin resistance which changes according to the reaction situation respective to the corresponding organ systems. And an acupuncture point value changed by a treatment corresponds to a change of the reaction situation of the organ system belonging to it.

Measurement of the acupuncture point values before and after a treatment permits a statement about the efficiency of the therapy. While

Electro-Acupuncture treatment according to Dr. Voll and related methods influence these specific skin zones from the outside with supplied defined electric stimuli in attenuated or stimulated sense, MORA BioResonance Therapy utilizes the body's own [*endogenic*] electromagnetic oscillations which are in a direct connection with the innumerable ongoing biochemical processes in the body.

These oscillations are measured on the skin surface of the patient, filtered in a physical-electronic system (MORA device), modulated and then given back to the patient as changed signals, with the directed goal of gaining a therapeutic effect through these changed signals in the area of the biochemical structure. Consequently, the previously measured pathological acupuncture point values return to normal.

MORA Diagnostics

MORA Diagnostics occur according to the basic principles of **Electro-Acupuncture According to Voll** (EAV). The acupuncture points react very sensitively and reproducibly with their measuring values when the patient's body comes in contact with compatible (therefore order promoting) or incompatible (order limiting) substrates. This way allows all kinds of substances like food, medications, allergens, parasites, toxins, heavy metals, among other things to be tested and to judge their influence on the body.

MORA BioResonance Therapy

MORA-Therapy represents a therapy very individually fitted to the patient by which his entire oscillation information is taken up by the device, is processed and is returned in a suitable form as therapeutic signals.

This individual treatment quite considerably favors the efficiency of this therapy form and allows access to many clinical pictures which are not easily accessible by less individual therapy forms. Additionally, tested oscillations of medications, colors, allergens, and nosodes among other things are included in the treatment as well.

MORA Allergy Diagnostics and MORA-Therapy

In an increasing trend, allergic illnesses belong to the most frequent chronic illnesses. Treatment in school [*orthodox*] medicine limits itself to the testing of suspicious substrates in the form of skin and inhalation tests with exclusively biochemical treatment forms like attenuated medications with corresponding side effects or very time consuming hypo-sensitizations (less effective and besides rich in side effects).

In school medicine the tested allergens are seen by their effect on the organism as equivalent with the result, that every allergen must be considered as individually as possible without inclusion of other allergens for the hypo-sensitizations.

In contrast to the ideas of school medicine which speak only of verifiable biochemical allergens, the idea of a cybernetically regulated organism gives a flowing transition between incompatible substances and allergens (antigens).

The incompatible substances are only physically verifiable with the typical allergic symptoms, and for unknown reasons have not led to an antibody formation with corresponding change of the immunology (RAST-Test).

On the other hand, the allergens (antigens) expel out additional biochemical changes (antibodies) by their physical verification.

Both groups can cause allergic symptoms in the same manner in the patient.

How important these exclusively physical verifiable intolerances (incompatibilities) are for later allergy treatment is shown by the fact that after a period of rest from these substances and a corresponding MORA BioResonance Therapy, bit by bit a change in the clinical picture with at first a subjective and then later an objective improvement of the allergy problems happens, without the specific antibodies in the RAST-Test changing. The treatment of the verifiable biochemical allergens alone is not sufficient in our experience.

Contrary to the evaluation of school medicine that looks at the discovered allergens as equally weighted in their effect it turns out that according to the experiences of biocybernetic medicine there is a hierarchy of allergens, therefore a different weightiness exists:

Ø Starting with a first allergen (causation or main allergen) bit by bit increasing sensitivities are joined with other allergens (trigger allergens) which then lead in total to a greater palette of incompatibilities/allergies, with the result that many patients finally react to practically all allergy causing substrates up to the everywhere appearing house dust, etc.

Ø Experience further shows that after finding the causative allergens and their careful elimination the other allergens as allergy triggers lose ever more meaning until finally the body no longer reacts incompatibly to them.

Therefore the allergy treatment decisively depends on finding and treating the specific causative allergen.

Obviously the disposition (heredity) and additionally the permanent supply of burdens – like certain basic foods, dental metals, environmental poisons, electro-smog, etc. – play a determining role in the sensitivity of the organism towards an incompatible/allergic substrate, which then more and more burden the control circulation system and lead beyond these false controls to the formation of further incompatibilities/allergies.

MORA Allergy Diagnostics

On the basis of research results with the medication test according to Dr. Voll, Dr. Franz Morell developed the allergy test within the scope of MORA BioResonance Therapy in 1985.

The procedure is based on the fact that instead of the real substance information (A-Mode = phase constant electronically amplified) the patient is supplied with the inverted information (Ai-Mode = phase constant electronically inverted) of the suspected intolerance/allergy substance.

On the basis of the burden the acupuncture measurement point would react to the pathological information (in A-Mode) with a measurement value deterioration. By inverting these substance oscillations it causes a measurement value improvement on the previously pathologically measured acupuncture points on account of the unburdening effect. Moreover, with this method the patient is not burdened by every single test, but is relieved so that there cannot be side effects.

Measurements of these material specific oscillations with different amplifications can capture a very elegant statement about the hierarchy and the degree of burdening of the tested materials on the organism.

In contrast to testing with biochemical preparations (skin and inhalation test) which need pharmaceutical preparations (factory preparations), with MORA testing all conceivable substances can be tested without immediate preparation. With it the therapist is not dependent on standardized test products and can introduce individually to the patient applicable foods, cosmetics, textiles and dental metals among other things directly in the test input.

The special value of the allergy test according to Morell lies in the fact that it represents a single test method for intolerance/allergy by which the body is not provoked, but is even relieved during the test process. The test methodology and the later treatment are practically painless and are easily practicable with children and sensitive adults. The possibility of the physical testing permits the examination of numerous substances in a very short time span.

MORA Allergy Therapy

After the intolerances are tested the treatment occurs. The discovered substrates are brought up to the patient with Ai (inverted) Mode with defined amplification and filter adjustments on the device. By the phase equivalents and modulated inverted allergen oscillations comes a reduction of the original allergen oscillations in the patient's body, which now optimizes his possibilities to eliminate the allergens resident in the tissues and as a result to decrease the sensitivity of the immune system to the allergy information.

Treatment with the tested allergen oscillations is carried out several times. Then another test of the discovered intolerances/allergens takes place to determine to what extent the treatments must be further continued – as long as an absolute period of rest from the tested remedies is complied with.
At this opportunity it must be observed that the MORA test and therapy method for allergy treatment described here works well alone, especially for children and young adults without additional chronic illnesses.

With older or multi-morbid patients, this test and treatment alone is not sufficient for lasting illness improvement. With these patients there always exists (besides the phenomena of the allergy) other restraining or even blocking influences of the organism which must be recognized and be switched off as much as possible.

Additionally, for example there are the constant burden from E-smog, incompatible dental metals, parasites or derailed acid-base household. Only then can a lasting treatment success be achieved.

Chronic diseases cannot be treated successfully by a simple "kitchen recipe at the push of a button" alone. The combination consisting of a MORA Allergy test and therapy as well as a treatment strategy which additionally supports the body and improves its general reactivity produces good results. For this constitutional therapy before and during the allergy treatment the MORA device with its big selection of treatment possibilities offers the best conditions.

The results of some case studies indicate that beyond the possibilities of school medicine the MORA BioResonance method can be a successful therapy.

YIN-YANG SYMBOL

This symbol represents the life force or Qi or Chi'I. Also positive and negative. The human body has Yin processing one side of the body and Yang the other. It also symbolizes the duality nature of man.

While the Chinese were probably the first to recognize the electromotive forces with in the body they discovered the art/science of acupuncture. By inserting very fine needles into any of 3500 known acupuncture points on the body they found that disrupting or connecting neurons can effect a positive or negative change to the patient. Stimulating these needles with the practitioners fingers the patient can verbally give feedback as to how they are affected, in a positive or negative way. Using these procedures on repeated visits have effected cures for many maladies that their patients have endured.

Electronic acupuncture does this non-invasively and sophisticated instrumentation provides a "closed loop system" allowing the practitioner to know immediately what issues the patient is suffering from and the remedies to eliminate those issues.

Most Practitioners performing Electronic Acupuncture (EAV) are Certified in German Biological Medicine. To simplify the "diagnoses" of major organs within the human body 50 points are tested on the appendages. Rather than seeking any or some of the 3500 acupuncture points that the Chinese use in their practice fingers and toes have electronic pathways to the major organs. These are probed and the readings from the instrument will alert the practitioner to abnormalities that exist.

Deviation from the norm is easily recognized and further testing can delve deeper into the affliction determining the kind and types of virus, bacteria or molds that are the culprit. More than 3000 Nosodes are programmed in the instrument and are identified in the patient using a comparator method. Once these pathogens are identified they can be eliminated, some immediately, others over a period of several treatments over several days. Rarely does any treatment last more than a month at which time all traces of the causal agents are destroyed and eliminated from the body.

Allopathic medicine nearly always treats symptoms, never the cause. EAV always attacks the cause, not the symptom. Therefore, the disease is always eliminated entirely using the methods attributed to German Biological Medicine techniques. These techniques use the body's own information to treat disease and goes directly to the source of that disease. This modality is preferred over any other method. Pharmaceuticals are rarely, if ever, used or required. Lifestyle changes such as changes in diet and even leaving your environment are always necessary to affect a lasting cure.

Electronics, electromotive and electromagnetism are all components of determining the health and wellbeing of an individual and are the most effective means by which to affect a cure. There is a mountain of data to support this fact and one need only to search it out to satisfy any doubter, especially Physicians.

A Word About Nosodes

Nosodes are a preparation of substances secreted in the course of a disease, also used in the treatment of that disease. Homeopathic practitioners use these to effect cures for the disease (Law of Similars, ie, like cures like, used by Samuel Hahnemann, M.D.) The Mora Super Instrument has 6,000 Nosodes programmed to facilitate the diagnoses and treatment of various diseases and pathogens. These would include bacteria, viruses, molds, allergens and much more! While homeopathic medicine is challenged by the allopathic medical doctors and FDA as spurious at best there is more than sufficient proof that, properly administered, homeopathic and acupuncture modalities are very effective!

WATER IS ALIVE!

Water is the most abundant substance on this planet. So natural and yet so mysterious. Even today the element of water presents a large group of researchers and scientists with many mysteries. Water is still a physical sensation. Even though this substance is so natural to us it is very exotic in its physical behavior. This extremely strange behavior cannot be explained with the formula H_2O. Water is more, much more than this formula can describe. Scientists discover almost every day unknown, strange and phenomenal properties of water. However, in very few cases that refers to chemically clean H_2O.

Water, as it exists in undisturbed natural surroundings, is completely different compared to technically clean water. It has many unusual properties of which a large number cannot be explained. In spite of the fact that science is still beginning to understand the subject of water, a great number of recorded scientific test results are available.

Combining all known research results of this exotic substance lead to only one conclusion:

WATER HAS MEMORY

As natural as water is for us, it is still very mysterious in its behavior.

Water molecules continuously form new structures when they absorb photon energy.

Water molecules are permanently in motion

Every indication in conscientious natural science research regarding the strange behavior of water, points to water having a biological memory. Like in a giant super-computer, nearly unlimited information can be stored in water molecules, then worked, manipulated and retrieved.

For a long time skeptics have rejected and attacked these new findings. Unfortunately that is what it is like in science but it basically confirms the speed at which advances are made. The intensity of the ignoramus gives a good indication regarding advances in better and more up to date developments. The whole history of science shows that even absolutely proven and demonstrated knowledge has great difficulty in being accepted by old fashioned mindsets, ignorant teaching methods and their advocates.

These are the so called "experts". A good example is the research done by Victor Schauberger. However, many of the skeptics have gone very quiet in the meantime. It seems that they have run out of arguments.

All these strange and already known properties of water are not a mystical phenomenon but represent the result of natural laws which by no means are all known or explained in science. However, that part which can be proven without any doubt is strange and astonishing enough to warrant more attention than it has received until now.

This still exotic substance is inextricably linked to all life on earth. In all organic life on earth water makes up a substantial part. The largest part of our physical existence is made up of this substance. Even our brain consists mainly of water (70-80%). However, this water is not to be compared to the lifeless technical H_2O.

Natural water is not a compact homogeneous substance as it appears to us to be but it is made up of a gigantic number of small and large molecule clumps.

Water researchers refer to clusters. These clusters are moving continuously. They therefore contain kinetic energy. This kinetic energy which is also known as "Brownian motion" is created by the absorption of light energy (photons). Water molecules are constantly vibrating and contain a substantial part of our life energy.

Depending on the amount of energy, water molecules form a variety of structures and as such behave accordingly.

Water molecules are constantly moving

Water molecules can take on various shapes. There are, however, particular ideal biological shapes of water clusters and these are of hexagonal shape. These hexagonal shapes are also present in snowflakes. In an ideal situation in the micro cosmos of the water molecules water clusters always form a hexagonal structure. Water researchers describe this as structured water. Structured water contains all the important bio energetic properties which distinguish lifeless technical water (tap water) from living hexagonal natural water. Hexagonal water is a kind of liquid life energy. Sophisticated laboratories continuously research the somewhat difficult to understand curious physical properties of water. One result is reproduced very time.

Natural water always consists of hexagonal structures.

Normal drinking water never shows this property as it is required to be sterile complying with hygiene regulations. This biologically lifeless water only produces chaotic diffuse lumps. In our efforts to understand water we have to accept additional effects beyond the chemical formula H_2O. At this stage it is not known which and how many. One thing is for certain and has been proven extensively beyond doubt in many bio physical research papers.

Water needs light and water stores light energy.

The top photo shows a frozen drop of structured water. Beneath is the picture of a frozen water drop from a communal water supply. The visual differences are very obvious. Water behaves equally different in all biological systems. Where structured water is present it stimulates all biological activities by transferring radiological energy of the absorbed photons. The water from the amorphous tap water ice lump is inactive and without life.

Depending on their energy uptake, water molecules build different structures which behave accordingly

Nature can restructure water
Fortunately nature is at any time capable to revitalize water. With the enormous power of the sun's energy, sufficient numbers of photons are transferred to the water. This starts the process of the **natural structuring of water**. Contrary to the light we produce with technical light systems, sun rays create the so called **"solar noise"**.

Nature delivers powerful energy impulses
Another important influence on natural water are the powerful energy impulses of natural electrical discharges. Extreme electric fields are created in the atmosphere by charging minute water molecules in the air and in the clouds. When discharging these powerful energy impulses, enormous forces are transferred to the water molecules. Water which is pumped up from underground sources does not contain this energy.

Natural magnetic fields support the restructuring of the water molecules

On our planet approximately seven thousand thunderstorms are active at any given time. The earth's magnetic field is affected by the intensive magnetic impulses created by lightning discharges. The natural magnetic field of our planet pulses within a narrow frequency range. Professor Schumann was the first researcher to systematically study these frequencies which are known as "Schumann Frequencies".

Of special importance of the Schumann Frequencies is a frequency known as "Global Frequency" with an average value of 7.81 Hz. This natural, biologically constantly present frequency is also used successfully in the medical magnetic field therapy. Important indications in water research confirm that water clusters respond favorably to this frequency field. Magnetic fields within this frequency band have an optimal influence on structuring water clusters. The magnetic field of the **UBS 315** unit pulsates at this frequency.

The systematic analysis of the most important and recorded physical phenomena of water from various sources show marked differences. Biologic life is always connected to the existence of natural water. In the past human settlements were always chosen because of the availability of natural water. Water which is nowadays easily available from the tap or plastic contains is not natural and is sure to spring some surprises.

Reprogramming water into its ideal biological condition.

The famous "Hunza Water" or fresh glacier water are well known examples of the extreme invigorating and vitalizing effect of structured water. Fortunately when researching **"Structured Water"** it was discovered that some of the important properties can be reproduced with suitable technical equipment. Based on the premise 'first understand it then copy it', today it is possible to reproduce "Structured Water". The important part of the required technology is a special photon irradiation (solar noise) and a weak pulsating magnetic field. Because of the absorption of light (photons) and magnetic fields hexagonal water structures are created from chaotic water clusters. Using available technology in a responsible manner completely natural energy fields can be created. Light from the socket and technical permanent magnets are totally useless for this purpose. The natural sunlight is made up of a completely different frequency spectrum (solar noise) than the technical light from the socket (100/120 Hz).

A good amount of intelligent electronic knowledge is required to produce natural identical fields.

A good amount of High Tech provides the solution

Energy and information can be transferred to water with suitable technical systems. Energized water has an enormous potential to optimize biological processes. It is this evolutionary biological water which has been available to all living beings for millions of years. Nowadays water is contained, stored or transported in technical equipment such as pipes and plastic containers and on top of that is changed chemically and mechanically and therefore does not contain these properties any more. In the long term this can substantially influence our wellbeing. Using suitable technology chaotic water can be reprogrammed to its ideal biological condition. An **U**ltra compact **B**ioenergetic **St**imulator completely incorporates the required properties in the best possible way.

A hand full of high tech makes it possible to give new biological vitality to drinking water

High tech makes it possible

Only a good measure of high tech is required to put new vitality into water. The latest in microelectronics, in combination with highly efficient LED's, change the water into an extremely energetic state by transferring it to the sun ray like "Solar Noise". At the same time the magnetic field modulated with the natural global frequency impacts on the water. In the same way the undisturbed power of the sun would in nature enhance water with new biological energy.

Modern high tech in an ultra compact format can restore water to the original biologically energetic condition

Regenerating water anywhere any time

A state of the art electronic instrument now available has two different photon outputs: white light and red light. The photon radiation is modulated with the so called "Solar Noise". White light during the day stimulates the production of endorphins and dopamine. Both are important hormones for health and wellbeing, fitness and awareness. Red light in the evening stimulates the sleep hormone melatonin. At the same time the red light (633 nanometer) has a very positive influence on revitalizing the eyes (macula, lymph, blood vessels, muscles).

The instrument also has an audio socket. The photon source and the magnetic field can therefore be modulated with any audio signal. This technique makes it possible to transfer subtle information to the water and into the subconscious (i.e. learning languages or musical water oscillations, etc.).

This instrument also has a scalar magnetic field. This is created by a planar coil built into the base plate. This type of coil is also known as a Tesla coil. Scalar magnetic fields have a considerable optimal influence on the creation of structured water. In addition to that the magnetic field pulsates at the so called "Global Frequency". The global frequency is also known as "Earth resonance frequency" (7.81 Hz). The alpha frequency (8-12 Hz) known as the secret frequency for super learning is very close to this secret frequency and optimizes all learning processes.

Energize water with this high tech miniature instrument.

Starting the day full of energy
All kinds of healthy teas whether it is herbal tea, green tea or white tea (which is the most beneficial) all develop a significantly better biological effect. This application is also very successful in improving the blood flow. Using a special blood test, dark field microscopy, the positive effect on the red platelets (erythrocytes) can be very graphically displayed.

It is important to drink water during the day
There have been many reports about the importance of drinking water. As mentioned on previous pages there are vast differences regarding the water qualities which are available to us. Not everyone is fortunate enough to live next to a natural spring. Very few can afford glacier water. Based on biological evolution humans are made exactly for this type of water. Nature does not know about water pipes or plastic bottles. However, with this instrument it is possible to restructure this biologically lifeless water. Structured water is a special biological power source.

Photons carry information
Photon penetrated water or other liquids carry the bioenergetic information before they are consumed. Bio energetic information can therefore be transferred to oneself without having to swallow anything. This is in particular a great advantage when bitter herbs or similar are involved. Like the ingredients of red wine (OPC) are active all the time while the glass is standing in front of you. Also crystals, or other effective information carriers like plant extracts dissolved in water, show remarkable results.

Structured water immediately improves the flow characteristics of blood

and other body fluids. Dark field microscopy confirms this.

Using dark field microscopy, the distribution and structure of the red platelets can be visibly reproduced at any time. Red platelets, not ideally distributed can have detrimental health consequences. Unfortunately with a large number of people who do not observe a healthy diet these blood conglomerates will look very similar.

Drinking structured water can bring immediate improvement

Using this instrument drinking water or any other drinking liquid (i.e. herbal tea) can be restructured to hexagonal water or liquid. Intensive photon radiation and the scalar magnetic field initiate an immediate structural improvement of the water cluster. This can immediately improve the flow values of the blood.

One glass of energized water can immediately improve your well being

Structured drinking water results in a cascade of positive effects on the flow ability on all body fluids. The clumping effect of the red platelets is dissolved within a short period of time and the blood shows a healthy viscose structure. An improved cell metabolism and the associated waste elimination is the result of taking structured water.

"God did not just drop us off" on the planet without giving us all we need to sustain, grow and propagate!

Déjà Vu
And the Thinking Man

From single celled Protozoa through higher animals each has the ability to fend for themselves from inception through death. Until we reach the mammals every species have instincts and inherent abilities that provide defense, finding and assimilating nourishment and reacting to various stimuli. These life forms do not require nurturing, teaching nor do they mimic others as they mature and live in their environment. Most proliferate in such abundance that, even though they are food for others, are subject to the vagaries of their environment or even have but a few moments of life spans, there are always sufficient numbers to procreate by any number of methods thus insuring the survival of the species.

While these biological animals have no abilities to think, reason, plan or create most have inherent attributes (given by God?) to hunt after a fashion, build clever traps or their abodes after a fashion, hide and even defend themselves by camouflage, sprays, speed and fangs. Their ability is usually unique to that species and is rarely duplicated in other species. Care of the young is entirely absent! These attributes are passed on from generation to generation through their own distinct DNA or Genes.

Mammals, while clearly at the top tier of living organisms in terms of intelligence, adaptability and nurturing their young seem to lack the basic survival instincts inherent in lower life forms. Mammals to survive must be taught by their parents or kindred the ways of food gathering, the hunt and how to protect or defend themselves from predators. This is usually accomplished by trial and error, repetitive reinforcement and real life experiences. Nearly all mammals require weaning and protection during infancy to assure survival of the brood.

Of all the mammals, with perhaps the exception of the porpoise and whale, only the primate has the ability to ponder, mimic and sometimes plan. While most mammals have instincts which govern much of their life, not so different from lower life forms, primates must be taught nearly everything that they know with the exception of reacting to stimuli such as heat, cold light, dark, pain and, most importantly emotion. Mammals left to their own devices at birth could very well perish from the simple rigors of just being alive. Other than finding the heat on their mother's bosom the primate must be taught to gather or hunt for food, avoid predators (unless they are one) and otherwise survive in a harsh world. In the mammal what is not instinctive is taught, what is taught is learned, what is learned is applied and what is applied assures the survival and procreation of the species. And so it goes, day after day and year after year.

The theory and notion propounded by Darwin and other evolutionists of selective breeding, survival of the fittest, species adapting by evolving their way from the primordial ooze, single celled entities communizing into molecular structures, sprouting appendages, crawling from the oceans and lakes to the land, growing wings to fly, longer beaks to delve deeper into flowers, growing hair to shield themselves from the cold and insects, brains that can think, spines, exoskeletons, multi-legs, no legs, two legged creatures is preposterous, entirely pseudo-science and just plain ignorance!

By simple observation and non-elegant logic it is clear that all species, plant and animal, from the lowest forms to the highest form of primate are part of a grand design, a grand orchestration of harmony, interaction, synergism and interdependence that obvious requires a Creator for this creation! Chaos is the natural order of all things. This can be proven in any number of ways. The most simple is the fact that there are no straight lines in nature or anything that is observable. Everything at the atomic level is moving, most of it at the speed of light, none of these move or form straight lines naturally. A higher intelligence had to make a straight line.

Heat will always tend to cool and will always cool over time. Our very own sun is cooling. It is the natural order of all things to be cool to its ambient. The natural order of all moving things is to slow down either by friction or depletion. Left unattended, roadways and walkways and other structures will give way to tenacious weeds, deterioration and oxidation, no matter how large, thick or long. These "destructive agents" can all be mitigated and changed by a higher intelligence by making order out of chaos! Order out of chaos is always a byproduct of higher intelligence!

The above posits are validated in The Second Law of Thermodynamics sometimes called Entropy. Science is mankind attempting to quantify the observable and unobservable. Science is quantifying what God has already created. Nearly all of "science" can be related in mathematical terms and, therefore, predictable. Even the chaos can be measured and quantified with some semblance of predictability. When anomalies or surprises occur in the scientific community it is given widespread attention with all sorts of predictions about its true impact on their science and even the future. Of course, these surprises are usually on the sub-atomic level or so far out in the universe that their "creation" or "discovery" had to employ accelerators and super powerful telescopes. That being a given, their impact on the world and the future are nil, at least in our lifetimes and, for nearly certain, tens of thousands of years if at all!

Mankind is the only species on the planet that regularly designs his own environment, not by instinct as with lower forms of life, but with clarity and of planned purpose. Mankind is the only species on the planet that can plan, deduce, have abstract thoughts, imagines, reasons, communicates with an enormous vocabulary to articulate every thought, desire, need and action. While other creatures may possess many of the same mannerisms, family unit, procreation method and primitive communication they are not able to achieve the heights that mankind has and will in the future.

The most intelligent of chimpanzees can be given every component of a watch or an automobile engine or a cuckoo clock and if they could live ten thousand years they would not be able to put any one of those together! Furthermore, the ability to mine, smelt, machine and weld and polish any of the components are beyond their capabilities. Written specifications, procedures and assembly instructions to repeat the processes beyond their intelligence to take raw materials to finished goods!

Scientists, anthropologists, bio-engineers and some other so-called experts proclaim to us that only 3% of our DNA separates us from the "monkey" or ape. Those claims lend credence to the evolutionists arguments how we are, therefore, descendants from the apes. While we wonder and are amused by the human like characteristics of the primates we are not to be swayed by the 97% which make us similar but should carefully scrutinize the 3% which differentiate us. These finite characteristics are simply an enhancement by God to cap off his masterful creation of man. It is apparent God has a sense of humor. By these similarities our Creator knew that mankind could be perplexed by this conundrum.

One only has to look at the Giraffe, Meerkat, African Gray Parrot, Manta Ray, Porpoise and other unique and fascinating creatures that roam the earth to realize that God very well has a sense of humor! Puppies, kittens and butterflies are an unending source of amusement to us humans and likely are to our Father!

If evolution is viable it would necessitate the existence of transitional beings all around us and readily identifiable. There would be amphibians with wings, birds with long skeletal tails, hairless apes that could talk (not to be confused with humans), lizards having gills, porpoises with feet and toes, humans with three eyes or arms (we could really use those) or a prehensile tail and……..well you get the idea. Evolution is a farce that cannot stand on its own "science!"

Humans certainly lack sensitivities endowed in other animals. Carrion and other birds of prey can see beyond the spectrum our capabilities. Canines can smell and hear far beyond our capabilities. Other attributes, such as agility, strength and flexibility are more pronounced in other primates far beyond our capabilities. Nevertheless, our intellect and how we are able to solve complex problems and think in the abstract are unsurpassed in the animal kingdom. Nothing can compare to our abilities to conceive, design, plan, execute and formulate our imaginations! No problem seems to be beyond our creative capabilities. Our innovations feed upon themselves to higher and greater advances. Our ability to record our activities, to store, sort and analyze data, to exponentially improve on technology establishing new and higher foundations from which to leap, combining a multitude of materials and ideas to bring about further innovation is the hallmark of modern man. These are God given attributes that defy all science and reason and yet they exist, some collectively and some individually.

The compulsion of man to search, explore and create is imbued by God our own Creator. And while we mere mortals cannot create something from nothing as our Most High Creator we do the have ability to mix, match and marry the created elements to suit our purposes. Our Creator did not give us steel sheets with which to fabricate products in various forms. Our Creator did not give us flour with which to make cakes, bread and pastry. Our Creator did not give us chemicals with which to clean, etch or lubricate. Our Creator did not give us mortar with which to make cement, seal the ground or bind bricks. Our Creator did not give us plastics or gasoline. Our Creator **did** give us the basic elements from which we have been able to innovate and create millions of items by combining these elements in innumerable ways. Our Creator created creators albeit more realistically defined as inventors, scientists and engineers! It is no stretch to believe that these were inspired and motivated by He who created!

Only man reaches out in search of or defining those things that are greater than him. Except for the most narcissistic who is not humbled by the forces he observes or the fear that comes from ignorance there is a need to worship a higher power as he comes to define it, from early man conjuring up gods for every celestial happening to images what these "gods" must look like. Until God actually manifested Himself to man called Adam and, much later, His peculiar and Chosen people Israel it was impossible to know Gods true nature and ultimate purpose for man. Thankfully the record exists in progressive revelation though His Word and the power of the Holy Spirit which works in the believers lives every day. No other faith, religion or cult realizes this God given power working in their lives on a daily basis very often so profound as to defy ones imagination.

Men and women especially, are blessed with something called intuition. This is a sense of the here and now and often of the unseen ramifications of a future event. This intuition does not seem to be brought about by any covert sensibility but seems to emanate from the intellect or spirit, an emotion, a feeling. Men sometimes get it in the form of a feeling of "fight or flee" even when no danger seems to be present.

How many times have you looked up from dinner or having a conversation with someone and noticed another looking directly at you? You weren't even conscious of that other person and he may have been standing or sitting across the room yet somehow you were directed to look that way and captured his look. It was a subconscious connection between you and he (or she) and it happens often enough in your life as not to be some accident of coincidence.

How many times have you been someplace where, even though it was impossible for you to have ever been there before, you just <u>know</u> you have been there before! What is the dynamic behind that thought or action?

A large number of people believe in reincarnation because they have a sense of living long ago or being another person from the past. A segment of the Jewish religion believes that the souls of the dead are in a repository to be reinstated in newborns. Still others "channel" to the dead thinking they can make a connection.

Auras are energy fields emanating from individuals which can be detected by certain photographic or electronic instruments. Most notable is that devised by Kirlian, a Russian scientist several decades ago. Kirlian showed that images created by Kirlian photography that depicts a conjectural energy field, or aura surrounding living things. Kirlian and his wife were convinced that their images showed a life force or energy field that reflected the physical and emotional states of their living subjects. They thought that these images could be used to diagnose illnesses. In 1961, they published their first article on the subject in the Russian Journal of Scientific and Applied Photography. Kilian's work was embraced by energy treatments practitioners and has been used with great success. **Many people are gifted** with the ability to see or sense these auras emanating from another individual. There are strong and weak auras which coincide with the personality and dynamic of the individual. Strong auras are usually attributed to Alpha personalities or spiritual strength. Auras exist and every living thing, including humans has one.

The fact is that the entire universe and every spec of matter is energy, which in order for life to exist at all there must not only be energy but harmony or harmonics. These are the "Good GOD Vibrations" that title this book. Every thought in the brain fires a synapse of energy converting one form of energy to another. This energy can traverse vast distances from as miniscule as one neuron to another across the room when one person senses another looking at them!

We are the sum total of our life experiences in what our conscious mind defines. However, we are the sum total of mans existence in the subconscious mind! Déjà vu, reincarnation or a sense of having done this or that, being there or seeing something is instinctive and an experience that we, to a lesser or greater degree have all possessed. Someone in our ancestral heritage did do that, did go or live there, did see that or something very similar and we were given a "flash" of that sense though our inherited DNA, Genome or Gene! Our chromosomes are packed with ancient information which we can only rarely tap into.

Throughout Caucasian history, never among the Blacks and often among the Oriental, a man presents himself in that era that surpasses all others in intellect, foresight, artistry and science. While there are thousands of such men the most notable that come to mind are Aristotle, Archimedes, Socrates, Pythagoras, Ptolemy, Tyche Brahe, Confucious, Michelangelo, De Vinci, Galileo, Copernicus, Newton, Beethoven, Bach, Mozart, Faraday, Tesla, Fermi, Bardeen, Shockley, Bessemer, Kaiser, Messerschmitt and a multitude of others. How were these icons of intellect able to tap into the otherwise unknown and thrust the civilization of the day toward immeasurable leaps? Some have suggested that they were genetically defective, autistic or bipolar. Others postulate that they had genius qualities that have set men apart from time immemorial. Others claim that they were visited by aliens from out there somewhere and bestowed these gifts.

It is my contention that these unique and gifted individuals were able to, knowingly or unknowingly, tap into the "collective" memory of mankind and collate this information, however sporadically, into a cohesive idea that turned into creativity and discovery beyond the capabilities of others. The great inventor, Faraday, who gave us foundational knowledge of electricity and the movement and storage thereof, would often ponder a problem for days. If he became vexed or frustrated enough he would isolate himself in a room without lights, sound or any other distraction. Except for water he would take no other sustenance. He would often remain for days only to emerge with the solution to his problem, without fail! He stated that the answer "would just come to him." Visitations? Not likely. How often have we gone to bed with a problem, a situation or conflict that sorely needed a resolution? And how often did the answer come in an epiphany in the middle of the night or upon awakening? More often than we might think.

The subconscious seems to work overtime when confronted with issues. If the conscious mind and day to day distractions can be held in abeyance the subconscious is free to correlate data from the here and now experience and subtly from our collective memories and formulate highly viable solutions to issues. Sometimes these solutions may seem to be so farfetched as to be unrealistic but they are solutions nonetheless.

It may be that certain drugs and even alcohol may suppress the conscious mind sufficiently to release the subconscious. There is ample evidence of this throughout history. Many great writers, musicians, scientists, philosophers and artists were known to imbibe, smoke opium or use other narcotics from time to time to achieve "breakthroughs" from an impasse. While the quantity and frequency of use and the measurable merits derived are questionable the results often are not.

We are the essence of God's creation by virtue of the intellect that He endowed us. Nothing on earth surpasses it. Nothing on earth compares to it. The notion and declarations that we are accident of nature and the resultant "virus" that are ruining the planet is a fabrication from those who loath themselves, mankind and, especially our Creator, God! An argument can certainly be made that man has embarked on a path of self destruction and annihilation. The Caucasian man is capable of copious amounts of destruction at the same time creating and inventing things that selflessly benefit all of mankind.

The creative power can be used for good and evil alike. What came about as a good thing can often be used for evil. A classic example of that is dynamite or TNT. While we can use it to clear land, carve mountains for roadways and railroad tracks and destroy obsolete buildings it can also be used as a weapon of war! Only the power of God and His Spirit within us will mans effort be for the good. Caucasian man has developed virtually all of the conveniences that the world uses for comfort, abundance and pleasure if they are astute enough to partake. Most of it was freely given and not hoarded for our own use or consumption.

We grow enough food for our friends and family and give or sell the remainder at fair and reasonable prices. We out produce everyone on the planet. We are the first of peoples to help others in need or after a catastrophe. No other group of people would ever help we Caucasians if we were in need or suffered a disaster. We can make the most complex of goods from raw materials from the ground. We can fly further, faster, higher and cheaper than anyone anywhere. Our homes are safe, clean, environmentally controlled and can last more than a hundred years. Our sanitation systems are unequaled in the world. We have 300 million citizens and an unknown number of non-documented illegal aliens on top of that in a country that is 3000 miles across and 1500 miles wide and we live in relative peace and harmony. This is about to change radically as Caucasians become marginalized, demeaned and ostracized. Violent crime is becoming the norm because of the infiltration of evil subversives in our once great nation. The powers that have stolen our country teach our children, feed us lies and distortions in their controlled media and send us to unwinnable wars for profit work diligently to remove all things of God from our psyche and culture and to subjugate us.

We will solve this dilemma the same way we have solved every other since time immemorial. It will not be pretty and it will not be nice but we will solve it when we have had enough, that is assured!

AN IMPORTANT BIBLE LESSON!

Your Preacher, Pastor and Teacher has no clue about the ultimate truth of the Bible, the things of God and, especially, the New Testament. The so-called Jews of today are not descendants from the Israelites of old and are certainly not "The Chosen" of God. Realizing this basic fact your entire perspective of the Bible and today's world will have a profound effect on your understanding of the Bible and your world view today.

The **Edomites** are descendants of Esau ("hairy, rough"), the eldest son of Isaac, and twin brother of Jacob, whose singular appearance at birth originated the name (Gen 25:25). Also, he was given the name of Edom ("red") from his conduct in connection with the red lentil "pottage" for which he sold his birthright (Gen. 25:30, Gen. 25:31).

Esau was much loved by his father and was, because he was first-born, his heir but was tricked into selling his birthright to his younger brother, Jacob, with the help of his mother, for a meal of red lentil pottage.*

(*Note: this Biblical version is rejected by the author. Much the same as Eve was seduced by Satan in the garden and bore the fraternal twins of Cain and Abel, Cain from Satan and Abel from Adam. Esau was a similar fraternal twin to Jacob as Cain was to Abel. Proof? Neither Cain nor Esau, although first borns were neve noted in the lineages as proffered in the Bible!)

Esau lost his father's birthright and his paternal blessing due to Jacob's subterfuge and thus raised the anger of Esau, who vows vengeance (Gen. 25:29-34; 27:1-41). Yahweh thereafter called Esau, "Edom" and the country subsequently settled by**Esau/Edom** and his brood was "the country of Edom" (Gen 32:3). Jacob, the grandson of Abraham and Sarah, the son of Isaac and Rebecca, is the ancestor of the Israelites. Later, Yahweh told Jacob that his name was no longer Jacob, but henceforth, Israel (Gen. 32:22-32; Gen. 35:10). Jacob's twelve sons were the ancestors of the **Twelve Tribes of Israel** and their descendants are described as the Twelve Tribes of Israel originally identified by the names of the twelve sons of Jacob: the Patriarchs Joseph, Judah, Issachar, Benjamin, Levi, Naphtali, Gad, Asher, Simeon, Dan, Zebulun, Reuben. Later, Joseph's two sons, his eldest, Manasseh and his second son, Ephraim, were adopted by Jacob as his own and so those two tribes replaced Joseph and Levi among the Twelve of Israel. The patriarch Judah was the fourth son born to Jacob (Gen 29:35).

The Edomites were thus the progeny of Esau, whose name was Edom, so called from the red lentil pottage he sold his birthright for to his brother Jacob (later, at Yahweh's behest, called Israel). These **Edomites** were also **separate** from the **Twelve Tribes of Israel** and so were *not* **true Israelites**.

They lived separately in a different land nurturing an enmity originating with their patriarch, **Esau/Edom** for **Jacob/Israel** and his descendants: a hatred born of a deep sense of injustice and betrayal that birthright and grace had been arrogated by trickery. Edom's violence against Israel (Jacob) was so intense not only due to a sense of betrayal but also because they both came from the same parents (Isaac and Rebekah); in fact, this great and enduring enmity began in Rebekah's womb, continued as the boys grew to manhood and endured until today in the phenomenon of the struggle of nations. Moreover, because of this enduring bitterness and jealousy, Satan's Esau would have destroyed Jacob had Yahweh not intervened

Ezekiel 35:1-15 describes Yahweh's judgement on, and devastation of, Edom who exulted over Israel's humiliation, who was their most bitter foe, and who "had a perpetual hatred to them, to the very name of an Israelite. "Esau/Edom in his hate and anger pursued "his brother with the sword, and did cast off all pity, and his anger did tear perpetually, and he kept his wrath forever." (Amos 1:11). Hence, the Edomites' "perpetual hatred" and "wrath forever" toward Israelites. That is, this seminal struggle of nations, which began in Rebekah's womb, endures today in the modern-day descendants of Israel (Jacob) and Satan's Edom

(Esau) .... a great struggle between the Israelites and the Edomites.

The Edomites lived and prospered in a land separate from Israel but were later attacked and defeated by Saul (1 Sam. 14:47) and some forty years later, by David (2 Sam. 8:13-14). Later, in the reign of Jehosaphat, (c 914 BC), the Edomites attempted to invade Israel, but failed (2 Chron. 20:22). They later joined with Nebuchadnezzar, the king of Chaldea, in his invasion of Judaea, the **Judaean kingdom of the Two Tribes**, and helped in his destruction of Jerusalem as well as the subsequent deportation of the Judaeans to Babylonia (c 630-562 BC).

The terrible cruelty displayed by the Edomites at this time provoked fearful denunciations by the later prophets (Isa 34:5-8; Isa 63:1-4; Jer 49:17). Afterwards, the Edomites invaded and held possession of the south of Palestine but they eventually fell under the growing Chaldean power (Jer 27:3, Jer 27:6). **The Edomites were thus Semites since they are closely related in blood and in language to the Israelites but they had no claim on the unique Bible Covenant and Birthright Promises gifted by Yahweh to Abraham, then to Jacob/Israel and then to his descendants**. However, for more than four centuries, the Edomites continued to prosper but during the warlike rule of the Maccabeans, they were again completely subdued, and even forced to conform to Jewish laws and rites, and submit to the government of Jewish prefects. Here, at this time, the Edomites become incorporated within the resurgent **Judaean kingdom**.

Edomites are therefore descended from **Edom** (Esau) whose descendants later intermarried with the Turks to produce a **Turco-Edomite mixture** which later became known as Khazars. That is, most of today's Jews are descendants of this interbreeding that produced a race called **Khazars** who had once governed an empire called Khazaria. Furthermore, this hybrid race **Edomite/Turk/Khazar** who created the Khazar kingdom and who between the seventh and ninth centuries AD, **adopted the religion of Judaism**. And, it is these **Khazar Jews** who are the ancestors of the vast majority of today's Jewish people. That is, **Edomite/Turk/Khazars** are the ancestors of the modern "Jews" including the Torah-true and Zionist Jews who spuriously claim right to the land of Palestine claiming it it is theirs by biblical demands and ancestral rights.

Consequently, the majority of today's Jewish people are known as "Jews" <u>**not**</u> because they are Judahites and descended from **Jacob/Israel** but because their **Edomite/Turk/Khazar** ancestors in their Kingdom of Khazaria **adopted the religion of Judaism,** called themselves **"Jews"** and **arrogated the Birthright Promises and Bible Covenants belonging to the Israelites, but especially those belonging to the Judahites.**

Thus, **"Jews"** are *not* **Israelites** and certainly they are not **Judahites.** Hence, modern Jews are *not* heir to the **Bible Covenants** nor to the ancient Nation of Israel given by Yahweh to the Israelites and the Judahites and so have no Divine Right or biblical mandate to the modern Land of Palestine.

Similarly, **Jesus of Nazareth** was *not* a **"Jew"** he was a **Judahite,** and **Jesus Christ** was *not* **"King of the Jews."**

So then, who are these people Israel, not to be confused with that land mass in the Middle East?

Flavius Josephus, in his ANTIQUITIES OF THE JEWS, says in Book XI, Chapter V, "Wherefore there are but two tribes in Asia and Europe subject to the Romans, while the ten tribes are beyond the Euphrates till now, and *are an immense multitude, and not to be estimated by numbers*." He wrote about the time of Christ.

In ISAIAH 62:2 God said to Israel, "Thou shalt be called by a new name which the mouth of the Lord shall name." HOSEA 2:17 reads, "They shall no more be remembered by their name." In ISAIAH 65 God said to the enemies of His servant Israel, "And ye shall leave your name for a curse unto My chosen: for the Lord GOD shall slay thee, AND CALL HIS SERVANTS BY ANOTHER NAME." (emphasis added) These "enemies" are yet called by their ancient name of "Jew," but Israel lost her name and identity, became the

"Caucasian" Race (from the mountains of our captivity), and now the "Christian" nations, named after the Name of our God and our Redeemer, the Lord Jesus Christ, FULFILLING HIS PROPHECIES TO ISRAEL!"

God brought over 40 million Caucasians to America (Zion) in one 50 year period, the greatest mass migration in all of human history, AND THEY HAD BORN AGAIN CHRISTIAN MINISTERS! Yet you preach the migration of a handful of atheistic and agnostic "Jews" to old Palestine, with their Rabbis who curse the Name of Jesus Christ, is "the fulfillment of the prophecies of the regathering of God's People Israel to their land!"

In PSALM 147 God said, "He sheweth His Word unto Jacob, His statutes and His judgements unto Israel, HE HATH NOT DEALT SO WITH ANY NATION AS FOR HIS JUDGEMENTS THEY HAVE NOT KNOWN THEM." The White Race prints and worships from millions of copies of "His statutes and His judgements:" and only Israel was to have them. Does this make God a liar; or is the White Nordic Race Israel?

The New Testament itself shows that all twelve tribes of Israel were well-known to Christ and the apostles in the first century. Let us start with the writings of James, the son of Joseph and Mary who was born after the birth of Christ. James knew exactly where the members of the twelve tribes of Israel were in the first century and he addressed his letter to them."**James, a servant of God and of the Lord Jesus Christ, to the twelve tribes scattered abroad, greeting"** (James 1:1).

The contents of James' letter tells us much about these twelves tribes of Israel. Instead of being wild and barbarous heathen tribes of Celts and Germanic peoples which history shows were in absolute heathenism at the time, these twelve tribes of James were attending synagogues (and synagogue services were conducted on the seventh day Sabbath) (James 2:2 -the word "assembly" in the King James Version is actually "synagogue," the official meeting place that Jews attended throughout the world and the KJV should have translated it that way).

Since James knew that all twelve tribes attended synagogues each Sabbath, it is no wonder they knew that Abraham was their father (James 2:21). They were well aware of "the perfect law of liberty" (the Mosaic law) (James 1:25) and James reminded them of what the Ten Commandments stated (James 2:8-12) They all knew the story of Rahab the harlot which is only found in the Old Testament (James 2:25); they knew of the story of Elijah (James 5:17); they knew what had happened to the patriarch Job (James 5:11); they were familiar with all of the Psalms of the Old Testament (James 5:13); they knew what the technical Hebrew term "Lord of Sabaoth" meant (James 5:4); and they were completely knowledgeable of all the teachings of the Old Testament prophets (James 5:10). Indeed, so familiar were these twelve tribes with "the scripture" (that is, the Old Testament) that James simply referred to the Holy Scripture as authority without once having to define it to those twelve tribes who were scattered away from Jerusalem (James 2:8). In fact, many of them had become "teachers" (the KJV has "masters") in matters concerning the scriptures (James 3:1).

Besides these things, James tells us that the majority of them were in types of business activities in which they traveled extensively from city to city (James 4:13). Their primary residences, however, were in regions that allowed James to use spiritual illustrations concerning fig and olive trees with which they were well familiar. In a word, James (who lived in the city of Jerusalem) knew where the twelve tribes of Israel were located in the first century and his letter to them shows they themselves were in constant touch with Jerusalem and the teachings of the Holy Scriptures. The apostle Paul was also quite knowledgeable of their whereabouts. When Paul was being tried in judgment before Festus and King Agrippa, he stated that he had lived the life of a strict Pharisee.**"And now I stand and am judged for the hope of the promise made of God unto our fathers: unto which promise our twelve tribes, instantly serving God day and night, hope to come"** (Acts 26:6,7).

Authors note: Paul was a Jew infiltrator to the Way of Jesus, the Christ. He suborned Jesus's words and doctrine and lied many times about who he was. Even though his writings occupy a preponderance of the New Testament the Church leaders and Academics have been in serious error about this subversive!

The peoples who made up the Celts, Angles, Saxons or Danes in the first century were our European ancestors in utter heathenism during the first century without the slightest knowledge of the Holy Scriptures. These, among the other "nations" of Europe were those to who the apostles were to bring the Good News, no others. Not the Black or Oriental!

The prophet Amos provided a major prophecy concerning the Northern Ten Tribes of Israel that has been overlooked by prophetic interpreters which fail to make the Celts, Angles, Saxons, Danes, etc. among the "lost" Ten Tribes of Israel. Amos said: **"I will sift the house of Israel among all nations, like as corn is sifted in a sieve, yet shall not ONE GRAIN fall upon the earth"** (Amos 9:9 Hebrew). It means that not one grain shall fall on the earth to germinate and to take root in the countries because they will still wander. While the prophecies show that many of these Israelites will live and die in the countries of their exile, the majority will never decide to give up their Israelite customs, to take root in those countries and remain there forever.

The Northern Ten Tribes will never return by the swarms to the place of their "roots." Jeremiah stated: "In those days the House of Judah shall walk with the House of Israel, and they shall come together in the land that I have given for an inheritance unto your fathers" (Jeremiah 3:18). And it is important to realize that the context of Jeremiah shows that the House of Israel and Judah will find their homeland in America!

The House of Judah and the House of Joseph (headed by the tribe of Ephraim -- ruler of the Northern Ten Tribes) will join together once again as one nation in a covenant relationship referred to by Zechariah with the word "Bands" (the banding together of a brotherhood union) and no longer will Israel be two divisions (Zechariah 11:7,14). Zechariah describes this covenant of brotherhood. **"I will strengthen the House of Judah, and I will save the House of Joseph, and I will bring them again to place them [in the Land of Canaan]; for I have mercy upon them and they shall be as though I had not cast them off: for I am Yahweh their God, and will hear them. And they of Ephraim shall be like a mighty man"** (Zechariah 10:6, 7).

Following the clues

To follow the history of the 12 tribes of Israel after the fall of their nation to the Assyrians in 721 B.C., we must recognize the path of their deportation and identify them by the names given them by their conquerors. Various websites and books have a great deal of information connecting the 12 tribes of Israel to the nations of Western Europe and the United States today, and it would be impossible to cover all this material in this answer. But here is some of the documentation.

When the Assyrians conquered Samaria, the capital of the northern kingdom, they transported many of the Israelites "to Assyria, and placed them in Halah and by the Habor, the River of Gozan, and in the cities of the Medes" (2 Kings 17:6). Shortly after the Israelites came into these lands, scholars note the appearance of peoples in this area called Cimmerians and Scythians. The Assyrians also called them Khumri, Ghomri, Gimiri (derivatives of King Omri of Israel) and Iskuza (derivative of Isaac).

The famous Black Obelisk in the British Museum includes a pictorial etching of King Jehu of Israel bowing and paying tribute to King Shalmaneser of

Assyria. The text speaks of Jehu, son (really a successor) of Omri, giving the Assyrian king silver, gold, a golden bowl, a golden vase, golden tumblers, golden buckets, tin, a staff and spears. This was the time during which Israel paid tribute to Assyria as a vassal nation prior to rebelling and being destroyed by Assyria.

Historian Tamara Rice writes: "The Scythians did not become a recognizable national entity much before the eighth century B.C. ... By the seventh century B.C. they had established themselves firmly in southern Russia. ... And analogous tribes, *possibly even related clans,* though politically entirely distinct and independent, were also centred on the Altai [Mountains of southern Russia and Mongolia]. ... Assyrian documents place their appearance there in the time of King Sargon (722-705 B.C.), a date which closely corresponds with that of the establishment of the first group of Scythians in southern Russia" (*The Scythians,* 1961, pp. 19-20, 44).

Boris Piotrovsky in his book *From the Lands of the Scythians* notes, "Two groups, Cimmerians and Scythians, seem to be referred to in Urartean and Assyrian texts, but it is not always clear whether the terms indicate two distinct peoples or simply mounted nomads. ... Beginning in the second half of the eighth century B.C., Assyrian sources refer to nomads identified as the Cimmerians; other Assyrian sources say these people were present in the land of the Mannai and in Cappadocia for a hundred years, and record their advances into Asia Minor and Egypt.

"The Assyrians used Cimmerians in their army as mercenaries; a legal document of 679 B.C. refers to an Assyrian 'commander of the Cimmerian regiment'; but in other Assyrian documents they are called *'the seed of runaways* who know neither vows to the gods nor oaths'" (1975, pp. 15, 18).

The Bible likewise indicates that the ancient Israelites would eventually migrate in a northwesterly direction away from Jerusalem. According to a prophecy yet to be fulfilled, God's Servant will "restore the preserved ones of Israel" (Isaiah 49:6), and these peoples will come from "the north and the west" back to Jerusalem (verse 12).While it is certainly clear that displaced Israelites were among these peoples, we should also note that not all Scythians or Cimmerians were Israelites. "Scythian" does not necessarily refer to a specific ethnic group. But it did include Israelites, who later moved in a northwesterly direction into Europe following their collapse as a nation.

Historians link the Cimmerians with the Gauls or Celts of northwest Europe

Historian Samuel Lysons linked some of the people who populated northwest Europe with these Cimmerians. As he put it, the Cimmerians seemed "to be the same people with the Gauls or Celts under a different name" (*Our British Ancestors: Who and What Were They?* 1865, p. 23).

English historian and scholar George Rawlinson wrote: "We have reasonable grounds for regarding the Gimirri, or Cimmerians, who first appeared on the confines of Assyria and Media in the seventh century B.C., and the Sacae of the Behistun Rock, nearly two centuries later, as identical with the Beth-Khumree of Samaria, or the Ten Tribes of the House of Israel" (noted in his translation of *History of Herodotus*, Book VII, p. 378).

Danish linguistic scholar Anne Kristensen concurs, stating: "There is scarcely reason, any longer, to doubt the exciting and verily astonishing assertion propounded by the students of the Ten Tribes that the Israelites deported from Bit Humria, of the House of 'Omri, are identical with the Gimirraja of the Assyrian sources. Everything indicates that Israelite deportees did not vanish from the picture but that, abroad, under new conditions, they continued to leave their mark on history" (*Who Were the Cimmerians, and Where Did They Come From? Sargon II, the Cimmerians, and Rusa I,* translated from the Danish by Jørgen Læssøe, The Royal Danish Academy of Sciences and Letters, No. 57, 1988, pp. 126-127).

The Bible likewise indicates that the ancient Israelites would eventually migrate in a northwesterly direction away from Jerusalem. According to a prophecy yet to be fulfilled, God's Servant will "restore the preserved ones of Israel" (Isaiah 49:6), and these peoples will come from "the north and the west" back to Jerusalem (verse 12).

Archaeological evidence

In addition to historical evidence, Scythian burial grounds have indicated a connection between these peoples and those of Nordic ancestry. For many years, scholars believed the Scythians were Mongols because groups of these nomadic people moved east, but the discovery of art and even a frozen corpse of a Scythian warrior indicate otherwise.

In July 2006 in the Altai Mountains of Mongolia near China and Russia, scientists made a rare find. German scientists who were part of the discovery team reported that the extremely well-preserved mummy of a Scythian warrior was that of "a 30-to-40 year-old man with blond hair" ("Ancient Mummy Found in Mongolia," Spiegel Online International, Aug. 25, 2006). Blond hair, of course, is a characteristic of Europeans not Mongols.

Prior to the discovery of this mummy, art obtained from numerous Scythian burial grounds had likewise indicated that these peoples were related to Europeans rather than Mongols. Because Scythian chiefs were buried with all their collected wealth, including wives, horses and art, detailed images of Scythians, their clothes and weapons have been uncovered. These discoveries depict their men with long, flowing locks, facial hair and Caucasian features.

In conclusion, biblical, historical and archaeological evidence indicates that descendants of the so-called 10 lost tribes of ancient Israel migrated to northwestern Europe. It is more commonly understood that many peoples from these nations also settled in the United States. For the above noted reasons, we believe that the peoples who settled in northwestern Europe and the United States are largely the descendants of the 12 tribes of Israel today.

More than a century before Darius the Great commisioned Behistun rock, however, the Assyrians were watching their borders. Their enemies the Urartians and the Medes may start gathering an army. The Assyrians wanted to know everthing that was going on with their foes. The border guards, and spies, would send back to the King, reports on the movements and activities occuring in the neighboring countries.

A good king always keeps a library of books and letters, etc. In 1847, Sir Henry Layard, uncovered the Assyrian capitol city of Ninevah. The Royal Palace contained over 23,000 clay tablets with everything from business deals to spy reports from the borders. It is these letters that become the transition point in prophecy fulfillment. Like Behistun Rock, these reports reveal the names used by the Assyrians for the different groups of Israelites they had placed as a buffer state between them and their enemies. With these names in hand we can trace these people and check the prophecies about them.

The "Royal Letters" are dated about 707 BC, only fourteen years after the fall of Samaria, the capitol city of the Kingdom of Israel. Maps showing the deportation and subsequent migrations of the Lost Tribes will show you the geographical locations.

Letters number 1079 and 197 were written by Sennacherib to his father, King Sargon. Letter 1079 tells of a resounding defeat of the army of the Urartians.

The troops were slain, and were fleeing. In the followup report of Letter 197 we find that this all happend in the land of Gamir. We still have one more step to go before we confirm that the Isrelites are the ones who live in that land.

In letter 112 it's reported that a people "went forth" from the midst of the Mannai, and into the "land of Urartu." Another Letter clearly separates the Urartians, the Mannai and the Gamera or Ga-me-ra-a-a as distinct from each other. The people named in Letter 112 are those Gamerraan; Cimmerians, in English. In captivity, the Israelites were renamed Gimira and Gamera and finally Cimmerians.

But these aren't ALL, just a smaller part of the total number of Israelites that were deported by the Assyrians. There were many thousands of others placed farther east. As the Persian writings on Behistun Rock show, these people were called by a different name. Among the prayer texts of Esarhaddon to the sun-god Shamash are several that name a people never heard of before in history, the "Iskuza" who evidently lived among the Mannai.

The name Iskuza can be easily deduced from the name "Isaac." The Israelites referred to themselves as House of Isaac before their exile - Amos 7:9,16. This name Isaac is the foundation for the name Scythian. Unlike the Assyrians, who gave the Israelites a name based back in the name of the Israelite King Omri, the Persians used the Israelites *own* name. History shows that the Iskuza were called "Shuthae" by the Greeks, and "Sacae" [also Saka and Sakka] by the Persians. Herodotus says that the Persians called the Sacae "Scythians." The Word Scythian only means nomad or wanderer, or one who lives in "booths." The word booth in Hebrew is Succoth, or scooth. The connections are obvious.

So, in the Assyrian tablets we are confronted with yet more Hard Evidence that the Israelites are not lost to history. Their names were changed. Why some of those folks have been traced to Japan! ! One of the most common Japanese names is Sakai. And their warriors have a name, Samurai, that is so close to the name of the Israelite capitol, Samaria, that the connection between Israel and Japan is virtually cemented.

In fact, the prophecy in Deuteronomy 33:17 that Israel would "push the people to the ends of the earth," is fulfilled. When one makes the connection between the Israelites and the Phoenicians, and, the Israelites and the Celts/Scythians, it's blantantly evident that the Isralites were the explorers/colonizers of all history. Those guys went *everywhere*. But that's another twenty subjects.

Here we zero in on the "transitional" names of the Israeites, allowing us to pick up their history, and thereby, reveal tens of prophecies which God has fulfilled. What better demonstration of God's effectiveness, than wielding some 3500 years of history!

Behistun Rock is one of the keys to finding the Lost Tribes. The crux is knowing the names used to identify those people. With the trilingual inscription of Behistun Rock we discover what three other cultures called the Ten Tribes. And it was the *Israelites*.

Behistun Rock is found in the Zargos mountains, in northwestern Iran, on an old caravan road that runs from Babylon to Ecbatana, the ancient capital of Media. The mountain is 1700 feet high and on the sheer face, 300 feet above the base is a huge bas relief commissioned by Darius the Great in 515 BC as a grandiose Ode to his great accomplishments. Listed are the nations and peoples he conquered and ruled as the king of the Medo-Persian empire.

The picture is accompanied by many large panels which are inscribed with three languages. The size of the whole monument is larger than half a football field; 100 feet high, 150 feet wide. One example of the quality of workmanship that went into the monument is the preparation of the surfaces. Where loose rocks and cracks were found, hot lead was added as a stabilizer or fill. At 300 plus feet! !

Sir Henry C. Rawlinson is mainly responsible for the decipherment of the inscriptions. It's interesting that Rawlison accomplished the feat of scaling the rock face while copying the inscriptions, and in 1840 deciphering the texts, all by the age of thirty!

The text contains many references that link Darius' subjects with the Israelites. The name "Kana", which is Canaan, appears 28 times. We also have a man named "Sarocus the Sacan who wears a Hebrew hat. Included in the nations listed is the Sakka. The term Sakka in Persian and Elamite becomes Gimri in Babylonian. Assyrian and Babylonian are often considered the same. We'll be hearing a lot more from the Gimri when we look at the Assyrian Tablets evidence.

On the monument you are able to see King Darius facing nine captives, which are secured by the neck with a rope. A tenth is under the King's foot. Each of these men is differently dressed. Across the bottom and up one side are many panels containing the story of Darius' conquests. There is also a large section of supplementary text.

The Behistun Rock inscriptions are confirmed in two other places: Darius' tomb, and a gold tablet. The gold tablet again mentions the conquering of the Sakka, while the tomb inscription expands the evidence by talking about three *different* kinds of Sakka. In all cases, the same name in Babylonian was Gimri.

Sakka comes from Isaac and becomes Saxon. Gimri comes from Khumri (out of the Biblical name Omri) and goes through Gimmira and the Greek Kimmerioi to Cimmerian. Almost all those names we learned in European history are traceable to the Sakka, Gimri and Scythians.

It's interesting that Darius was putting down insurrection among the Israelites, while he was assisting the Israelites in rebuilding the Temple. Just the reverse happened a centruy earlier. Some mercenaries from the House of Israel came and helped Nebuchadnezzor in the siege of Jerusalem. Archeologists have found the typical three sided arrow heads used by Israel in a city gate they uncovered.

After being taken captive and relocated below the Black and Caspian Seas, the tribes of the House of Israel plus tens of thousands of Jews were used by their Assyrian conquerors as a buffer state to ward off any advances by the Medes. Soon, groups of Israelites started moving out to east, and north. The main body of people remained in the area for about a hundred years, during which time they fought as mercenaries for just about everyone. Their unique triangular arrow points were even found in the ruins of one of the burned gates of Jerusalem; meaning that some of them were in on the conquering of Jerusalem by Nebuchadnezzor! !

Soon after the power of the Assyrians was broken, vast numbers of the Israelites began several migrations, with the main two groups moving west under the Black Sea, and north through the Dariel pass of the Caucasas mountains into the steppes of south Russia. A large group also moved east. These were called Sakka (Saka) and Iskuza by the Medes and Persians. They headed west to populate northwest Europe and the far western island of Britain. Some of these migrations were undertaken all the way into the 17th century when, in the final migration, some from the tribe of Manasseh sailed the north Atlantic to Plymouth to fulfill Isaiah 49:20, which was prophesied to the hegemony of the House of Israel, Ephraim (England).

You see, the crux of the whole LT subject is that the names we want to look for to trace the Israelites are not the names that historical accounts and archeological finds give those same folks. To more confuse the issue, large groups of Israelites called themselves by different names. Some of them called themselves the House of Isaac, which is pronounced e-*sahk* with the emphasis on the last syllable. How natural for the Persians to call them the Sakka(Sacae in Greek), while the Assyrians called others, the House of Omri, after the sixth king of Israel. This name sounded like Khumri, and was variously pronounced Ghumri, Gimri, Gimira, Gammer, all of which turned into the Greek Kimmeroii, our English word Cimmerians.

The Israelites weren't lost, their *name* got lost. That fact coupled with the **erroneous search for the *Jews'*** fulfillment of the Old Testament prophesies has held the Lost Tribes teaching in virtual obscurity these millenia since 500 BC.

732-700 B C
Israel taken into exile by the Assyrians who called them Khumri, later corrupted to Gimira.

710-590 B C
Israelites, called Gimira by the Assyrians and Kimmeroii (Cimmerians) by the Greeks, established a reign of terror in Asia Minor. They finally migrated to Europe, to a place which they called Arsareth (2 Esdras 13:40-44 of the Apochrypha)

600-500- B C
Following the collapse of their Assyrian allies, the Scythians were driven north through the Caucasus by the Medes, and they settled in south Russia.

650-500 B C
Cimmerians in Europe moved up the Danube and became known as Celts; the English derivative of the Greek Keltoi.

525-300 B C
Others driven out of south Russia by the Scythians moved north-west between the rivers Oder and Vistula to the Baltic, where they later became known as Cimbri.

400-100 B C
The Celtic expansion from central Europe: some attacked Rome in 390 BC and settled for 200 years in northern Italy; others known as Galatians, after

invading Greece in 279 BC, migrated to Asia Minor. Most of them moved west into France and later to Britain.

250-100 B C
South Russia was invaded from the east by the Sarmatians, who drove the Scythians north-west through Poland into Germany.

A D 450-1100
The Romans re-named the Scythians Germans ("genuine") to distinguish from the newly arrived Sarmatians in Scythia. Some of these came to Britain as Anglo-Saxons, AD 450-600; others, after moving north through Jutland, became known as Danes and Vikings. Some of these came directly to England, but others settled for a short time in France and were called Normans.

THE NAME'S THE GAME

This is an incomplete list of names for various groups of Israelites as they migrated.

Khumri, Cimmerians, Sakkas, Sacasene, Sacasune, Schythians, Cimbri, Thraco-Cimmerians, Celt, Galatians, Germans, Saxons, Normans, Danes, Gimira, Kimmeroii, Iskuza, Gauls, Angles, Picts, Iberes, Scots, Basques, Bretons, Goths, Vandals, Lombards, Franks, Burgundians, Ostrogoths, Daci, Belgae, Massagetae.

God makes two kinds of promises in the Bible, conditional and UN-conditional. He promised Israel blessings for keeping His Law. He promised Jeroboam a dynasty for keeping His Law. These were conditional promises. However, God gave no conditions when He told Abraham that he would be the father of many nations, and kings would come out of him. Gen 17:4,6. Likewise when He said that the heir would come out of Sarah. God put no conditions on His word. He spoke it as though it would be fact. This is also the case when God talks about David in II Sam 7:16.

David is promised, unconditionally, that his bloodline would never run out. God also said that David's kingdom AND throne would go on forever. David had no say in the matter. He couldn't do anything to hasten or stop those events from occurring. God didn't give David a choice, He just said that those things would come to pass. The whole passage from verse 10 on doesn't seem to make any sense.

When this prophecy of Nathan was given, Israel was at peace with everyone. In fact David ruled from the Euphrates to the Nile. There was peace, prosperity. The Israelites had finally settled down in one place and united as one nation under God and David. How can Nathan say that God has appointed a place for Israel, He would plant them IN that place, from which they'd move no more, and be free from their old enemies? Weren't they IN their own place? And who was thinking of moving? We'll come back to these questions. Verses 11 through 13 don't look so unusual. They portray David's soon to come family history. David will be "made an house (dynasty), he will die a natural death, Solomon is to be king, Solomon will build the temple. And the follow-up is easily disconnected from David. It says that "his (Solomon's) throne shall be stablished forever." BUT, the first and last parts are very hard to accept. Well, the part about David's bloodline (his house) might be OK. David had a lot of kids and they all lived 3000 years ago, so there are probably some people around who come from David's bloodline. But can anyone believe that the last part is true? Does David's "throne" exist today? Today is part of forever, isn't it?

Go one step further in establishing the impossibility of this promise. We have to tie these two parts together. Psalms, Jeremiah and I Kings all say that the "house" and the "throne" go together. I Kings 2:4, "shall not fail thee a man on the throne of Israel." Did you notice how that statement adds a whole new dimension to the picture? Not only must there be a direct, bloodline descendant of David ruling, he or she must be ruling over some Israelites!! Is it possible that in today's modern world we can fulfill these conditions? Yes, but only through the teaching on the Lost Tribes.

Fill out this promise to David by looking at Psalm 89. There are two renderings of the events. First, an overview of the basics, verse 4. David's "seed is stablished forever, and his throne to all generations." Then starting in verse 22, David is promised that no enemy would exact (prosper, Isaiah 54:17) upon him, nor through war would his old enemies, Canaanites, Egyptians, Edomites, etc., hurt him. He also gets honor and wealth and power.

Then God reiterates the forever part in verse 29. Not satisfied, God makes sure we heard by pinning His promise to the Sun and the Moon. Jeremiah echoes Nathan in chapters 31 and 33. In 33:17 he says that, "David shall never want a man to sit upon the throne of the house of Israel." And 31:35-36 says if there's a sun, moon or stars, there'll be a nation Israel.

The only conclusion that can be drawn from these promises is that somewhere in the world, this minute, there is a bloodline descendant of King David ruling over the descendants of the Israelites.

Let's show off God a bit. See how He worked out this mission. Gen 49 is a crossroads of the Bible. Yet most people don't understand the implications of Jacob's actions in passing out the blessings and birthright before he dies. Jacob gives Joseph, and therefore his boys Ephraim and Manasseh, all the good stuff. Virtually the whole birthright. Jacob only keeps back the right to be king of the family and make it's governing laws. This means that although Joseph and his family will be the rich nobility of Israel, someone from Judah(a Jew) will be king.

When Joseph dies, the Tribe of Ephraim will be the pre-eminent tribe of the nation, but will accede to Judah in being king. As soon as Joseph dies in Egypt, someone from the tribe of Judah will take the rulership under Pharaoh. This person will be from the Zarah family of Judah; he was firstborn. Two hundred years after the secession of Judah from the nation Israel, Assyria conquers Samaria and deports the ten tribes(millions of them) to northern Mesopotamia. Most of them migrate to northwest Europe and the British Isles over the next 17 centuries. But never fear, we still have a Judahite king in Jerusalem named Zedekiah. Here comes Nebuchadnezzor. He conquers Judah, kills Zedekiah's sons and carries Zedekiah and 95% of the people to Babylon. The line is cut!! What about David's promise now?

The Zarah line is ruling all across the northern Mediterranean. By this time they've got colonies in Spain, Ireland, and Britain. But the "throne of David" is not in those places. It is in Jerusalem. And the male bloodline has been cut! But not the FEMALE. God allowed the daughters to inherit when no male heirs were available. So Jeremiah takes the King's two daughters, and leaves for Ireland. He drops off one daughter in Spain and then marries the Pharez-Judah daughter of Zedekiah to the Zarah-Judah High King of Ireland. This final move sets things straight; the way God outlined them. Zarah is back on top. The "throne" is preserved. The resting place for Israel is being prepared for the arrival of the ten tribes as they begin to migrate across Europe.

All things are possible with God.

There is a progression of the Royal lines Zarah-Judah and Pharez-Judah. Queen Victoria spent one million Pounds establishing her genealogy. There is a wonderful genealogy chart produced by R.H. Milner which lists ALL the kings to descend out of Judah. Here is the list of Royal lines included on this chart: Troy, Sicambrians, Franks, ancient British, Byzantine, Tudor, Anglo-Saxons, Norse, Normans, Guelph, Wettin, Skiold, Irish, Scottish, and Hungarian.

TROJANS/MILESIANS/CELTS/SCYTHIANS

There is one place in Spain that is named after a Biblical family; however not one of the tribes. This is Zaragosa, which comes from Zarah, Judah's first-born.

It was Zarah's family that ruled for over a 100 years in Egypt after Joseph's death, and before the Egyptians revolted against the (Hyksos) foreign rulers. At that time the nobility, the Zarahites, fled across the Mediterranean sea to already established cities. Cities that had been founded by their own family: Troy, Athens, Miletus, and more. Diodorus talks of two exoduses out of Egypt, one by sea, one under Moses. These Zarahites then migrated west, the Milesian/Zarahites arriving in Spain and Ireland near 600-700 BC.

One famous Trojan/Zarahite named Brutus landed in Britain 1103 BC and founded New Troy; later to be renamed London.

It's this Zarah stream of the Judah line that makes sense out of Ezekiel's prophecy of 17:22-24 and 21:25-27. The high tree is the Pharez line, the low tree is the Zarah line. Him that is high is Pharez, him that is low is Zarah. It was the Pharez line that came out of Egypt as the rulers of the Judah line(Jacob gave Judah the sceptre and lawmaking). With the Zarahites gone to the northern Mediterranean, the Pharezites took their rightful place as rulers.

The exalting of the Zarah line was when Jeremiah married off one of King Zedekiah's daughters to the High Zarahite King of Ireland about 583 BC. Thereby bringing the two lines together and establishing the Zarah line again as pre-eminent. All the kings and queens of the British Isles and northwest Europe come out of that union.

Didn't God say that David would "never want for a man on the throne?" These kings and queens out of the line of Judah were and have been ruling over the Lost Tribes these many centuries: 25-plus.

George Washington and other American Fathers had ties back to British nobility. And John F. Kennedy's heritage included kings of Ireland

THE MEETING PLACE OF JUDAH:

ZARAH -- PHAREZ

God had to get the "Isles afar off" ready for the House Israel, so He sent two Houses of Zarah-Judah; the Trojans to Britain and the Milesians to Ireland. Brutus, the grandson of the Trojan King Aeneas, rounded up hundreds of defeated rebels after the siege of Troy, and made his way to Spain where some of his countrymen had previously migrated. Many of these compatriots sailed with Brutus as he went to Britain. He landed there in 1103 BC, moved inland somewhat until coming to a likely spot for a settlement. He gave this town the name of New Troy. This was later to become the capitol of the reconstituted tribe of Ephraim, and the final resting place of the Davidic Throne. London. God was also preparing this land for the influx of Israelites which started some 500 years later.

Some centuries later, God had replaced those Spanish Zarahite Trojans with their cousins the Zarahite Milesians. After living for some time in Spain, some of the princes struck out to establish new kingdoms in Ireland. To do this they had to conquer their cousins of the tribe of Dan; the Tuatha de Dannan. After a family squabble that left only one of the original three brothers who came to Ireland, Eochaidh (Yo-kee) The Heremon (king) assumed the throne of Ard-Righ, High King. This all happened around 600 BC. Eochaidh was well in place to receive the bloodline of the Davidic line of Pharez-Judah.

Within three years of the final fall of Jerusalem, God's Trustee of the Davidic Throne, arrives in Ireland with his scribe Simon Brug, a curious collection of ancient relics, which include a harp, a large chest, a three hundred pound stone, and more. Beside one other in the party, the band of four was rounded out by the daughter of the blinded, imprisoned King of Judah, Zedekiah.

It wasn't enough for God to keep the promise that Judah would supply the rulers of the Tribes of Israel. If that were the case, then God could have quit right there and had a day off. He would have had some Zarah-Judah kings ready for the Lost Tribe's arrival starting around 500 BC. But He'd made that promise to David. So David's bloodline had to be in the mix, too. That's why He had to draft <u>Jeremiah</u> to "plant" that bloodline in the future land of Israel. God's plan for Israel had been side-tracked. With the marriage of Zarah-Judah Eochaidh to David-Pharez-Judah Tea Tephi, things were again set right according to God's Plan, Promise and Prophecy:

    1-Judah would have the sceptre; Gen 49:10
    2-David would have his "man on the throne;" II Sam 7:16
    3-Israel would assume it's new land, II Sam 7:10, to "renew" it's strength, Is 41:1, *for the spreading of Christianity.*
    4-The fulfillment of a multitude of prophecies by Ezekiel, Isaiah, Hosea, and others.

Jeremiah even set the future "priesthood" of the church in motion when he established Colleges in Ireland. He was the precursor author of our Constitution. Besides schools, he's the great Law-Giver of Ireland. It's Jeremiah's Law that became Irish Common Law, then British Common Law, which processed through our Pilgrims as the Mayflower Compact, and became our Constitution. Jeremiah also instituted the Three Great Faires of Ireland. Do you suppose that these *political/business* gatherings fell approximately at the same times as Passover, Pentecost, and Tabernacles?

Knowledge is Power. God *knows* everything, hence is All Powerful.

Who have been the two most powerful nations on earth, excluding that Johnny Come Lately the USSR? The United States, as a single nation. Britain as an Imperial nation. These two countries controlled most of the world's information. They are the ones who "pushed the people's to the ends of the earth." They have for 200 years controlled and contained most of the world's knowledge.

The point I make is that we're well educated people, we Americans and British. That condition is no accident of history. It's a move on God's part. **2500** years ago God sent Jeremiah to Ireland to establish the educational system that would weather the Satanic Ignorance of the Dark Ages, and provide a foundation for our modern level of education. Who still has the best schools today? The Irish.

Or perhaps you think the various conquerors and emperors of history, and even the mighty Roman Empire itself, had a blind-fit and never knew that Ireland was there. The Romans went to Britain and spent many resources to maintain a hold in Britain. They never could bring the whole country in line, as Isaiah predicted, Is 54:17; they had to give up and build walls. They wouldn't have had near the trouble with Ireland.

Ireland is a VERY special place. It's the Cradle of the Spiritual and State governments of modern Israel.

The ties between Jesus and Britain are many. The traditions surrounding Glastonbury make it certain that Jesus spent much time there. God has been "grooming" Britain all through history. From the arrival of Brutus in 1103 BC, Jeremiah in Ireland in 583 BC, the Druid (Hebrew) Priests, the Megalith Builders of prehistory, and the endless waves of Celts and Scythians that migrated to that area, I think we can conclude that Britain has a part in God's plan.

In another place, there is evidence of the family ties of the British Royalty to the Roman Royalty. The story reads like a Payton Place of History. Inter-marriages and intrigue. Love and War. Good and Evil. The Roman Emperor who ended centuries of Christian persecution, and declared Christianity the religion of the state, Constantine the Great, was born and raised, in Britain. And was half British, on his mother's side.

There is a quite God-typical connection to those Royal folks. Joseph of Arimathea, who was Mary's uncle. He was Jesus' great-uncle *and* guardian after Joseph the carpenter died; somewhat early in Jesus' life.

Not only is *this* connection to Jesus' family wonderful. See just who it **was** that made up that family.

Looking at it humorously, Joseph of Arimathea's daughter married an earthly king and his sisters married God. Anna, married King Beli, and produced the lines of Queen Elizabeth, King Arthur, and the Royal family of the Roman church. Ann, one of Joseph's sisters who became the Great Matriarch of Jesus immediate family, was great-grandmother to "half" the Apostles! ! ! AND Bianca his other sister became grandmother to BOTH Jesus and John the Baptist. No wonder he jumped inside Elisabeth's womb. His **cousin**, the Messiah, had come to visit.

JEREMIAH'S VOYAGE

An old man arrives on an Island with a small group of people in 583 BC. He brings the daughter of a King, a scribe named Simon Brug and some relics. The powerful Milesian High King of all Ireland allows the old man complete control. Instituting laws, schools and congresses, the old man forever changes the face of the Island's history, and subsequently the history of the entire world. Apparently incidental to all this, is the fulfilling of a 500 year old prophecy.

Few people know that Jeremiah was much more than a prophet. He tends to get lumped in with Isaiah, Ezekial, and the others. Jeremiah did more than go around speaking doom and gloom. He held a high level position in the kingdom of Judah. He was the grandfather of King Zedekiah. II Kings 24:18. Most importantly, Jeremiah was God's Trustee of the Bloodline and the Throne of David.

Jeremiah's commission has always puzzled scholars. One can find where Jeremiah rooted out, pulled down, destroyed, and threw down kingdoms. History shows that his prophecies about the destruction of kingdoms came true. The mystery is, where did Jeremiah "build and plant?" The scriptural account doesn't contain any building and planting. There is also some confusion about Jeremiah's being put "over the nations." It would appear at first glance that this meant his prophesying against them. This is not the case. First, Jer 1:10 says that God set him "over *the* nations, not nations (in general). This is repeated with the word kingdoms; *the* kingdoms. The bible is concerned with only one people, the twelve tribes of Israelites. Jeremiah was to "throw down" *AND* "build and plant" *the* Israelite nations. We'll have to follow his trail to find where he accomplished his mission.

First, look at the Biblical account. Jer 15:11-14 tells us Jerry is going to a brand new place he "knowest not." Isaiah fills out the picture a bit. Isaiah 41:1-3 tells us that a "righteous man from the east" was put over nations and kings. This man would not travel by foot (on land). Jer 41:10 establishes the presence of the "king's daughters" in the group with Jeremiah. Jeremiah, as their great-grandfather, would certainly have assumed the postion of Guardian.Then we find Jeremiah and the girls going to the Egyptian city of Taphanhes. In fact, there is an ancient structure there that bears the name, "Palace of the Israelites daughters." Isaiah helps us again with a last bit of confirmation, in chapter 37:31, telling us that a "remnant of Judah" shall escape and "take root downward."

Before going on, we must take notice of what God had promised Jeremiah and his fellow travelers. God told Jeremiah that he'd be treated kindly by the Babylonians and die a natural death. Baruch, Jeremiah's scribe and Ebed-Melech, the Ethiopian, are also told they'd be spared. The probable number in Jeremiah's traveling band was five: Jeremiah, Baruch, Ebed-Melech Tea Tephi and her sister.

It's not so hard to trace the migration of large groups of people. Not so with small groups. But God knows this too, and has left evidence that we may overcome our doubts about Jeremiah's destination. But we have to go the history books. Only one place in the world claims to have the grave of the prophet Jeremiah. Only one country's history tells of an old man, and his scribe Brug bringing a king's daughter from Egypt. Only one country claims the Harp of David for it's Arms. Only one country has Jerrys coming out of it's ears. **IRELAND!**

Although, due to the Bards embellishing the story, accounts of Jeremiah's arrival and work in Ireland differ in some details, the basic elements of each tale are the same.

> The Stone, known as the "Stone of Destiny" came from Spain, and before that, from Egypt
>
> It came in the company of an aged guardian, who was called "Ollam Folla", (Hebrew for revealer or prophet)
>
> Accompanying the man was an eastern king's daughter
>
> Eochaidh (Eremhon) married the daughter, Tea Tephi
>
> The aged guardian became the most influencial Statesman and Spiritual leader of Ireland.

Here is the evidence that God would supply us to confirm Jeremiah's trip. There is an inscription found in a tomb located in Schiabhla-Cailliche, near Oldcastle, County, Meath, Ireland, not far from Tara. Thirty-some stones with strange markings upon them, lie in the sepulchral chamber within the huge cairn of stones which make up the tomb. A large carved stone outside the tomb is still pointed out as Jeremiah's judicial seat. Our confirmation lies on those thirty stones in the cairn.

One interperation, by George Dansie of Bristol, says that the stones show a Lunar Eclipse, in the constellation of Taurus and a conjunction of the planets Saturn and Jupiter in Virgo. The prow of a ship is shown in the center, with five lines indicating the number of passengers it carries. On the left, a part of the ship, perhaps the stern, is shown with only four passengers, one having been left behind, as indicated by the line falling away from the ship. The wavy line indicates the passage of the ship across the ocean, terminating at a central point on an island.

The stellar and planetary alignment of the inscription gives a date of 583 BC. This date allows just the right amount of time for our little band to go to

Egypt, and return to Palestine briefly before making their way to Spain, then Ireland.

TEA TEPHI

Buried ineradicably in the poetry and folk-lore of Ireland is the tale of a Prophet, an Egyptian Princess and Simon Brug (Baruch) a Scribe. They Landed in Ireland about the same time that the destruction of Jerusalem took place, bearing with them a great chest and a stone wrapped on a banner. The Princess married the Zarahite King, Eochaidh II. Ard-dath, Ard-righ, or Heremon (horse man of all Ireland), and their son was Irial. I, (M.R. Munro Faure) give quotations from old Irish verse:

The praises of Tea Tephi, daughter of Lughaidh (equivalent in Erse of Bethel) are sung as:
"The Beautiful One with a Royal Prosperous Smile."
"Tephi (Hebrew beautiful) the most beautiful that traversed the Plain."
"Temor of Bregia, whence so called."

Relate to me O learned Sages,
When was the place called Temor?
Was it in the time of Parthalon of battles?
Or at the first arrival of Caesaire?

Tell me in which of these invasions
Did the place have the name of Tea-mor?
O Tuan, O generous Finchadh,
O Dubhan, Ye venerable Five

Whence was acquired the name of Te-mor?
Until the coming of the agreeable Teah
The wife of Heremon of noble aspect.
A Rampart was raised around her house
For Teah the daughter of Lughaidh (God's House)

She was buried outside in her mound
And from her it was named Tea-muir.
Cathair, Crofin not inapplicable.
Was its name among the Tuatha-de-Danaan
Until the coming of Tea - the Just
Wife of Heremon of the noble aspect?
A wall was raised around her house

For Tea the daughter of Lughaidh,
(And) she was interred in her wall outside,
So that from her is Tea-mor.

A habitation which was a Dun (Hebrew court) and a fortress
Which was the glory of murs without demolition,
On which the monument of Tea after her death,
So that it was an addition to her dowry.
The humble Heremon had

A woman in beautiful confinement
Who received from him everything she wished for.
He gave her whatever he promised,
Bregatea a meritorious abode

(Where lies) The grave, which is the great Mergech (Hebrew burial place)
The burial place which was not violated.
The daughter of Pharaoh of many champions
Tephi, the most beautiful that traversed the Plain.

She gave a name to her fair cahir,
The woman with the prosperous royal smile,
Mur-Tephi where the assembly met.
It is not a mystery to be said
A Mur (was raised) over Tephi I have heard.
Strength this, without contempt,
Which great proud Queen have formed

The length, breadth of the house of Tephi,
Sixty feet without weakness
As Prophets and Druids have seen.

From "Forward" - Watchman What of the Dawn

WHO OWNS THE ROCK?

The coronation stone which sat in Westminster Abbey, England, is the coronation stone of the Hebrew nation called Israelites. This stone was named Beth-el (house of God) by the patriarch Israel (sometimes called Jacob) roughly 2000 BC and remained with his descendants. It travelled with them for forty years in the wilderness, supplying their water, and was preserved and brought to Ireland in 583 BC by the prophet

Jeremiah; eventually being transferred to Scotland, then England, and now resides in Scotland.

STONE HISTORY

In 1950 BC a man lays his head on a rock, has a dream of God, and calls the rock "House of God." His family carries that rock around for the next 38 centuries. It becomes their symbolic throne, their water supply, the type of their coming Deliverer, their coronation symbol, and it even roars when the proper king is crowned.

> Gen 28:10-14 Jacob puts his pillows on a rock, dreams of a ladder to heaven.
> Gen 28:18-22 He sets up the rock for a pillar (of witness), annoints it with oil, naming it "Beth-el"; Jacob also names the place Beth-el (House of God)
> Gen 31:13 God validates Jacob's "Bethel" name and annointing.
> Gen 35:9-15 Jacob's name is changed to Israel, God reconfirms His promises, Jacob again annoints and names the place Bethel.
> Gen 49:24 The Stone of Israel (Jacob); the Shepherd is passed on to Joseph
> Josh 24:24-27 Joshua sets up the Stone at Shechem for a witness.
> I Kin 12:1 Rehoboam goes to Shechem (in Ephraim) to be crowned king of Judah.
> I Kin 12:25 Shechem is established as the capitol of the kingdom of Israel.
> II Kin 11:13-14(II Chr 23:13) Joash is crowned by a pillar, "as the manner was."
> II Kin 23:3(II Chr 34:31) Josiah's covenant by a pillar.
> Jud 9:1,6 Abimelech is Gideon's son; he is crowned by the pillar at Shechem.
> Hos 3:4 Israel is to be without a standing pillar.

REJECTION

> Coronation stone rejected for Solomon's temple--Ency. Freemasonry, 1921
> Mt 21:42; Mk 12:10; Lu 20:17 Jesus quotes Psalm 118
> Ps 118:22 The stone which the builders refused......
> Acts 4:11 Peter says Jesus was the stone refused
> I Cor 10:4 "And did all drink the same spiritual drink: for they drank of

that spiritual Rock that went with them: and that Rock was Christ."

JACOB RE-CAP

He dreams at Moriah. He annoints and names the stone Bethel. Later God confirms Jacob's annointing and naming. Jacob calls the stone the Shepherd of Israel and gives Joseph the birthright; Gen 49:24.
    Who gets all Jacob's fine jewelry?----- Joseph.
    Who gets Jacob's treasured ointments?----- Joseph
    Who gets all Jacob's revered relics?----- Joseph
    Who gets Beth-el?----- Joseph
    What tribe retains all Jacob's good stuff when Joseph dies?----- Ephraim
    To whom, logically, will the places named Bethel belong?----- Ephraim
    Josh 16:1-2; 5-8 Mt. Bethel belongs to the tribe of Ephraim

THE DESERT TRIP

OVERVIEW: Exodus under Moses, 1453 BC. Multitude of 2.5-3.25 million. A hurried departure. At Sinai, the Ten Commandments. Refusal to act in faith at Kadesh-Barnea. Forty years in the wilderness. Joshua is made leader. Conquest of Canaan.

    Deut 29:5 Shoes hath not waxen old upon thy foot
    Ex 16:35 children of Israel did eat manna forty years
    Ex 15:23-25; :27 bitter waters turned sweet at Marah; 12 wells at Elim
    Ex 17:1-6 at Horeb/Rephidim. God stands on *the* rock. Before all, Moses strikes the rock once and all drink from the issuing waters
    Num 20:7-11 at Kadesh, Moses strikes the rock *twice*
    Num 20:17-19 Moses writes a letter to Edom asking passage through the land, without drinking water.
    Num 21:21-22 Moses writes same to Sihon, King of the Ammorites; at Arnon
    Location and size of Edom and Ammon: below and northeast of the Dead Sea, roughly 75 miles thourgh each country.

THE ROCK

ALL references are to a specific rock. Not "a" rock, not "some" rock, not "rock", but THE rock.
    Ex 17:1-6 God stands on THE rock
    Num 20:7-11 THE rock(four times) at tabernacle door.
    Deut 8:15 THE rock of flint

Deut 32:13-15 Honey/oil of THE rock
Neh 9:15 water from THE rock
Ps 78:15-16 streams out of THE rock
Ps 78:20 smote THE rock
Ps 81:16 honey out of THE rock
Ps 105:41 he opened THE rock
Ps 114:8 turned THE rock
Is 48:21 clave THE rock (twice)
I Cor 10:4 spiritual Rock; THAT Rock was Christ

THE GROOVE AND THE ROAR

There is a groove worn deeply into the Rock between the two metal rings. This would have to result from many years of carrying the Rock on a pole. And this had to have happened before 583 BC when Jeremiah brought the Rock to Ireland, because it has been moved a very few miles in the last 2500 years. This groove is the result of being carried around the wilderness for forty years.
Irish legend says that any imposter or unrightful heir to the throne would be known by the fact that the stone would roar ONLY when the rightful king stood on it.

TWO KINGDOMS, TWO HOUSES, TWO ROCKS

Birthright
House of Israel: Joseph gets Bethel Rock, Bethel mountain, 10 Tribed kingdom

Sceptre
House of Judah: Judah gets the Spiritual Rock, Jesus.
Both Rocks were rejected. Bethel so it could travel to be with The People, Israel, in the British Isles.

The Coronation Chair of England has been in constant use to crown the Monarchs of England since 1296 AD when Edward I had it constructed for his coronation. The chair was built specifically to house the Coronation Stone which Edward brought from Scotland. It had resided there since being brought by Fergus from Ireland in 500 AD. It's earliest use as a Coronation stone in Ireland was 583 BC when Eochaide the Heremon was crowned High King of Ireland after his marriage to Tea Tephi, the daughter of Zedekiah, the king of Judah conquered by Nebuchadnezzar.

All the kings of Israel, the whole nation, then the Kingdom of Judah, were crowned standing on or beside this Stone. See the story of Queen Athaliah's overthrow in II Kings 11:14. Upon Jeremiah's arrival in Ireland, 583 BC, with the Stone, the Stone was again put in use to crown the Royalty of Israel. Eochaide and Tea Tephi were the complete fulfillment of Jacob's command that Judah should rule over the people of the twelve tribes.

Eochaide was a descendent of the Zarah line of Judah, while Tea Tephi was of the Pharez line of Judah. The Zarah line, as firstborn, ruled in Egypt after the death of Joseph. Non-ruling princes of the family migrated over the next 150 years, founding the city-states of Troy, Athens, Miltetus, and others along the northern Mediterranean. Then about 150 years before the Exodus, when the King of Upper (southern) Egpyt conquered the Hyksos(Zarahite) rulers, the ruling family fled leaving the Pharez line of Judah in Egypt. This "first" Exodus out of Egypt is mentioned by the historian Diodorus.

Brutus of Troy went in 1103 BC and founded New Troy, later renamed London. The Milesians migrated west establishing settlements in Spain and then Ireland. With the arrival of Tea Tephi, and her subsequent marriage to the Milesian king Eochaide, the two lines of Zarah and Pharez were brought together. From this first marraige, all the kings and queens of northwest Europe would descend.

Notice that the three times the Stone moved is the fulfillment of Ezekiel's prophecy regarding the three overturns mentioned in chapter 21, verses 25-27. However, now that the Stone has been moved back to Scotland, some have discounted the prophecy. This mistake is made because the Stone has been equated with the Throne of David. The Stone is only a symbol of the Throne. The actual Throne still remains with Queen Elizabeth. The Throne will remain with Britain until Jesus returns, according to the prophecy; "until he comes whose right it is."

The latest descendant of the Zarah/Pharez line of Judah is Queen Elizabeth II. She also happens to be the current fulfillment of the prophecy of Nathan to King David in II Samuel 7:10-17. In that passage, David is promised that he would have a direct bloodline descendent ruling over the people of Israel forever. Psalm 89 confirms that this would be the case as long as the sun and moon can be seen in the sky.

The Great Seal *of the* **United States**

Heraldry is a very interesting subject, but when applied to the Lost tribes material it can be amazing. Most of the symbols used by the tribes of Jacob's descendents are derived from the blessings and prophecies that were given to his sons in Genesis 49. For instance, the Wolf is the primary symbol of the tribe of Benjamin (verse 27, "Benjamin shall raven,["eat voraciously"] as a wolf). Reuben's two symbols are a Man and Water. Judah's symbols are a single Lion and three Lions respectively. These may be resting or rampant.

Although Joseph's two sons Ephraim and Manasseh are not listed, we know from other sources that Ephraim and Manasseh each had two symbols. Ephraim used the Ox primarily and the Unicorn was his secondary symbol. Manasseh's primary symbol is an Olive Branch, while his second symbol is a Bundle of Arrows. It must be remembered the Manasseh is the thirteenth tribe. Ephraim moved into Joseph's place and Manasseh was added to the twelve. If one gives the briefest look at the arms of England, it will be seen to contain a shield with a rampant Lion on the left and a Unicorn on the right. Other symbols such as David's Harp are included within the shield.

The United States, being founded as a permanent home by the Pilgrim's (of the tribe of Manasseh), should reflect the emblems of that tribe. It will be found that the Great Seal of the United States contains ten features of thirteen. And given that some sixteen drafts, over a time period of 159 years through which the seal was constructed, it makes coincidence impossible as an explanation for those 13's.

It is possible that all these thirteens is that there were thirteen original colonies, but this surface explanation doesn't fit all the evidence. When the facts of the United States and the Great Seal are viewed in the context of the whole Lost Tribes teaching, it obviously fits like it belonged. The "coincidence" of there being thirteen colonies doesn't add a whit to the total picture unless it is taken to be part of the LT material. Finally in 1935 the final die was cut for the seal that we see on the one dollar bill today.

The obverse, or front, side of the Seal as proposed by Thomas Jefferson was to show the Israelites being led by the Pillar of Cloud and of Fire. Benjamin Franklin suggested the event of Pharaoh's drowning in the Red Sea. The Harp of David was included in three early designs, while the number of olives and leaves did not settle to thirteen each until late in the nineteenth century.

Although Congress had appropriated the funds (1884) for the "obverse and reverse of the seal of the United States", the Reverse Seal was not cut. Five times the "acts of Congress" relating to the Reverse Seal were ignored; 1782,

1789, 1833, 1884, and 1902. It took 153 years for the complete process!

E. Raymond Capt, in his booklet, "Our Great Seal" says: The Great Seal given to our country, after years of laborious heraldic and symbolic study, reveals our true national origin and destiny. The Obverse face is Israel in the Old Testament; The Reverse face is our race under the New Covenant. Each face is a masterly harmony of all that is potent in symbolism and prophecy. It was originated and adopted by men who recognized the overshadowing presence of the Great Architect of the Universe and submission to His will as revealed in the Scriptures and the Laws of Nature. They planned a government in conformity to His great Plan. They recognized that America's greatest task was to go toward the goal of the Plan -- the eventual establishment of the Kingdom of God on earth.

Their hands were guided by another, for seemingly they did not fully know themselves to be of the Tribe of Manasseh. ("Blindness in part is happened to Israel" - - Romans 11:25). Yet, wittingly or unwittingly, they used all the national emblems of ancient Israel as America's emblems also. Our Obverse Seal sums up the whole of the Old World history of Israel, whose boundaries were set at the very beginning according to the number of the sons of Jacob, and the Lord as an "Eagle" was his express guardian and "Shield." Unsealed "Truth," borne by the Eagle's beak illuminated the way of the "people saved by the Lord." The blessings and the curses were plainly set before them in the "Olive-Branch" and the "Bundle of Arrows." The Constellation of Stars, signifies perpetual endurance. The Almight God himself swore by them as such. (Daniel 12:3)

Isn't it interesting that there is a most definite connection between God's People and the Great Pyramid of Giza?

13'S FOUND IN THE GREAT SEAL
    13 stars in the crest
    13 stripes in the shield
    13 olive leaves
    13 olives
    13 arrows (some like to include the 13 feathers of the arrows)
    13 letters in Annuit Coeptis
    13 letters in E Pluribus Unum
    13 courses of stone in the Pyramid
    13 X 9 dots in the divisions around the crest.

Get out a fairly fresh dollar bill and a powerful magnifying glass and see Israel in Our Great Seal.

Although the evidence is nearly two hundred years old, that makes up only 8% of the total years from the fall of Samaria in 721 BC. With the publishing of the 29th chapter of Acts in 1800, the first clue was manifest. It tells how Paul heard that Israelites were living in Britain, and determined to go there. This ties in with Jesus' admonition to go *first* to the "lost sheep of (not Israel, not the House of Judah) the House of Israel; the ten tribes. If you trace the Celts and Scythians, you'll find that those are the ones Paul went to; after the Spirit refused to let him go to Asia.

Another facet: The 7 times punishment period decreed by God in Leviticus 26:18, "if ye shall NOT keep my statutes," ran out for Ephraim (England) 2520 years after the fall of Samaria, 721 BC. Research shows one time to be a "year of years." A 360 day year. 360 X 7 =2520. No other time period applied to the term "time" will yield a satisfactory period for the promise to be fulfilled. And the book of Hosea is a virtual History of what happened to the Celts/Scythians.

As a little exercise, you could assume that the Celts/Scythians are the House of Israel, then trace an over view of their history; to the present. They end up just like Hosea says, being "known as sons of the living God." These people, with NO knowledge of their heritage, turn out to be the only other people on earth besides the Judahites who are connected to the Bible and It's God. You'd think, on the surface, that the Arabs (being from Abraham) would be more likely to be the same. They aren't!

So God carefully concealed the evidence spoken of above until His punishment period had run. Then notice in the mid-eighteen hundreds how archeology exploded. Layard dug up Ninevah, to uncover the Royal Library of Ashurbanipal. Those tablets give us a lock on the NEW names to look for when searching out the Israelites. Likewise, Rawlinson, at roughly the same time, disciphered Behistun Rock, and the Assyrian tablets were confirmed. Darius II also left confirmation of the Israelites new names two places in his tomb.

There is only one hurdle to cross when considering the Lost Tribes: find that transition time in history. The Lost Tribes have a historical Pivot Point. This is the point at which the Israelites become the Gamir (variously spelled) and the Sakka. We have sure record of the Israelites before this point in time, and we have sure record of the Gamir and Sakka after this time. So once we establish the link between the Isrelites, the Gamir(Celts), and the Sakka(Scythians), we've got mountains of the evidence.

One other small point: you know how slow science/history/archeology is. It's only a little over 150 years since Layard and Rawlinson. Just look at the Dead Sea Scrolls. It's over forty years and the public still doesn't have it all.

Evidence, evidence. We understand evidence very well. We do almost nothing in our lives without evidence to back it up. Admittedly, some of us choose faulty or bad evidence, but we DO have evidence. Use the example of buying a house. A lot of Christians gather more evidence when buying a house than they do to confirm God's reality or workings. Of course we wouldn't call them Christians, but the churches are full of folks like that. They can't answer two serious questions about their belief system.

A researcher named Barry Fell has written three volumes on the archeology of North America, particularly the US. He proves, without a shadow of a doubt that Israelites were here as early as 1000 BC! There's a four foot square rock face in New Mexico, that contains the Ten Commandments almost verbatim! From the time of Christ, just outside Moneta (money), Wyoming, there is a rock face inscribed for trading, with the name of the bank included. It says in effect, "1st Iberian bank, no usury." Does "no usury" sound familiar? Doesn't sound like much 'till one knows that Iberia was settled by the descendants of Eber, the Hebrews (not Jews). The words Iberia and Hebrew are cognates. Same goes for Hibernia (Ireland).

God was sending Israel, as Isaiah says, to the "isles afar off" to "renew their strength" in order to take the Gospel to the world; which was their job from the beginning.

THE SYMBOLS OF ISRAEL

Tribe	Sign	Color	Stone	Became	Emblem	Waiting
Rueben	Aquarius	Orange	Odem	France	Man/Water	Behold a son
Simeon	Pisces	Red/Orange	Saphire	Silurians (Senones)	Sword/Gate	A hearing
Levi	Pisces	Red/Orange	Saphire	(Judahites)		Joining
Judah	Leo	Blue	Nophek	Juti (Judahhites)	Lion/3 Lions	Praise

Zebulun	Capricorn	Yellow/Or.	Pitdah-Topaz	Holland	Ship	A habitation
Issachar	Cancer	Blue/Violet	Bareket	Swiss/Dutch/Fins	Laden Ass	My hire
Dan	Scorpio	Yellow/Gr	Leshem	Danes/Swedes/Norw.	Serpent/Horse	Judge
Gad	Aries	Red	Yahalam/Ruby	Sweden	Troop	Good fortune
Asher	Libra	Green	Shebo	Scots/Germans (orig)	Cup	Happy

Naphtali	Virgo	Blue-Green	Yashpeh (Jasper)	Norwegians	Stag	My wrestling
Benjamin	Gemini	Purple	Achelamah	Iceland, Norway Normans	Wolf	Son of right hand
Ephraim	Taurus	Red-Purple	Tarshish	Britain	Ox, Unicorn	double fruit
Manasseh	Sagittarius	Yellow	Shoham/Onyx	United States	Olive Branch, Arrows	Fogetting

Z a r a h					Red Lion, Red Hand	Rising

The Bill of Divorce

The issue of Israel's divorce is an important one to address. It is one of the keys to identifying the Lost Tribes, and keeping them separate from the Jews (House of Judah).

Jeremiah tells us about God divorcing the House of Israel. And Isaiah gives it a confirming side comment. This is followed up by Hosea's statement that Israel is "not my wife".

820 JER 3:8 divorce of Israel
802 IS 50:1 moms bill of divorcement
977 HO 2:2 Israel not my wife

Israel's idolatrous ways lead to God giving them a "bill of divorcement." But that created an obstacle between God and the people of the House of Israel. By God's law, a woman or man who was divorced had to wait until the partner died before they could remarry. Otherwise the person was guilty of adultery. Since God gives us no indication that that law has been changed, we must go on His last statement about one partner dying.

We need to pause here and do a little study. The law of divorce is given by Moses in Deuteronomy 24:1-4. But you won't see anything about not being able to marry in that passage. Quite the opposite. It describes re-marrying clearly, saying that the woman given a bill of divorce may marry another husband. The first husband may never take her back; even if the second husband divorces her, or if he dies.

But the divorce law was re-interpreted by Jesus, just as he redefined the law of murder. In Matthew, chapters 5 *and* 19, Jesus says that the man who divorces a woman and takes another, makes the first woman commit adultery. This is adequately witnessed again in Mark 10:11-12, but adds that the man commits adultery too.

Again, in these New Testament passages, there is no mention about one partner dying. But please consider, does the man whose wife dies commit adultery by marrying again? It's pretty hard to commit adultery against a dead person. So if one of the divorced partners dies, the charge of adultery can't be brought.

One last question before we return to the main topic. Is it possible for God to commit adultery? He can't break His own law. Therefore, before the House of Israel could re-marry without committing adultery, legally, her first husband had to die.

In the form of Jesus, God died. This freed the House of Israel from the bond of the law, and allowed the establishment of a connection between the formerly divorced house and God.

Isaiah, starting especially in chapter 40+ is an explanation of the coming messiah and his relationship to the House of Israel. This account is a more detailed account of the same events described by Hosea. When reading these chapters of Isaiah, keeping reminding yourself that the House of Israel is England, the US and northwest Europe. The activities and prophecies will make perfect sense to one who knows a bit of the history of those people.

By the time we get to Isaiah chapter 54, we find the divorced wife being referred to as the "barren one."

This divorcement issue also helps make sense of something that Jesus says. He was "not sent but for the lost sheep of the House of Israel." He could have easily said what the fundamentalist Christians like to think, that he was sent for the lost sheep of the world. Another common Christian misconception is that Christ came for the Jews, but, fortunately for the Christians, the Jews rejected Jesus, opening the door to the rest of the world; especially them. Other interpretations of Jesus' "mission" are equally faulty.

Jesus said "house of Israel" because he referred to his task of reuniting the divorced House of Israel and God. Certainly the Jews didn't need reuniting with God. They were still known as the people of the God of the Bible. And even a beginning student of scripture knows that the Jews were of the Tribe of Judah, specifically, and as an adjunct, the tribes of Levi and Benjamin (with some from Simeon thrown in).

Jesus even gives adequate witness by sending his disciples to those very same House of Israel folks. The Lost Tribes study will show that a majority of the locations visited by the apostles were those on the westward route taken by the Lost Tribes on their march to their new "place"; that Nathan prophesied to David in II Sam 7.

So, while there are numerous references to the Church being the Bride of Christ, this has no bearing on the Lost Tribes until it is recognized that the overall concept is the House of Israel; which means more than those who are "saved." Salvation is a side issue, not the main issue.

Jesus came to re-establish contact for an express reason, which was NOT the salvation of the House of Israel. It was so that they could continue their God-given task of informing the rest of the world about His coming Kingdom. You'll note that Jesus didn't preach salvation. He preached the kingdom of God. Look it up.

God's purpose for the Israelites, all twelve tribes, is to get the word of His kingdom out to the rest of the world. The Jews fumbled the ball and became a "closed" society. They were never initiatory in disclosing God to those around them. They didn't sent out missionaries as did England, the US and other European nations.

Those nations have been recognized by the rest of the world as people who profess the God of the bible. They have become what Hosea said they would, "known as the sons of the living God."

There is a possibility about the first period of Hyksos rule, which is founded in very little evidence, that evidence being mostly non-existent, is that the first Hyksos ruler ship of Egypt, well before Abraham's time, was headed by Shem after he defeated Ham, and eventually went to Palestine to found Salem. The book of Jasher states that Abraham was taken by Terah to live with Noah and Shem for the 10th-49th years of his life. This is where he learned the things he taught the Egyptians.

The second period of Hyksos rule was from Joseph, spanning as much as 250 years until the Pharoah of Southern Egypt "rebelled" and indentured the leftovers who didn't go out in the first exodus; namely the "tribe" of Zarah as leaders over many groups of immigrants to the north of Egypt. It is already well documented that Darda and Calco, sons of Zarah, founded and ruled settlements/kingdoms in what would later be called Greece and Troy.

The terminus of this migration is very natural because during the 250 year rule of Joseph-Zarah in Egypt, the tribe of Dan pioneered all of Europe; naming all the rivers and other geographical benchmarks after himself, as was his wont. The briefest study of the rivers of east and west Europe will turn up many names that contain some form of the word Dan; Dan ube.

The Danite-Phoenicians had to scout out the route to be taken by the Lost Tribes on their trek to the place prophesied to David by Nathan in II Sam. The place where their traditional enemies would bother them no more.

Hyksos rule was carried out by the Adamic Priesthood and the Israelites. The last Adamic High Priest Shem ruled the first Hyksos period, and the Israelite/Hebrew sons of Jacob; Joseph, Zarah, et al, ruled the second.

Do not be confused by the name Hyksos. It's a word like Scythian. It's only a description, not the name of the described. Scythian basically means nomad. They went around in succoths/scooths/scyths.

Hyksos only means "foreign." Foreign kings. Some like to say "Shepherd" kings. There weren't a people known as the Hyksos. This is very easy to understand when applying it to the second period of Hyksos rule. They were Israelites, but to the Egyptians they were Foreign kings, Hyksos. Don't we get our version of the history of that time from the people who kept the records of that time? The Egyptians? They called the Israelites what they wanted. And they hated the Israelite rule.

This is one of the reasons that there is so little mention of the Israelites in history. It's as though (and some have offered this) the Exodus never occurred and those millions of folks are a product of imagination. I've heard of new ruler ship going around tearing down all the monuments/references to the hated regime they replaced. It is no stretch to see the Egyptians doing the same.

There is plenty of evidence in the corners that paints the true picture. It's just very hard to come by all in one place. One must take in such disparate sources as Gordon, Sitkin and Velikovsky.

The Study of the Lost Tribes is the study of the history of the earth. The Israelites virtually took over the planet. From Japan going west to California.

That sounds rather broad brushed, but consider. What was God accomplishing with all of us Israelites? What is His ultimate goal for humanity? The "trees" are the Lost Tribes. The Forest is the total history of the Earth, as regards humanity. We have to back way away from the trees to see the Forest, no?

God's objective is to bring the knowledge of Himself and His coming earthly kingdom to all of His Chosen on the planet. What better way than to lead groups of Israelites to all the corners. They will be the most receptive to God's call; when He chooses to wake them out of the "spritual sleep" of worshipping/recognizing other than Himself.

God works in the long term. *Centuries* before an event He'll send out the scouts, establish travel routes, prepare destinations for habitation, send rulers who will "instantly" embrace the "message/way"of those who come later. Brutus to found New Troy (London), Heremon to Ireland for the arrival of Tea. Esther, Tobit, etc. We accept this readily on the small scale of Joseph being "sent" to Egypt to prepare a place for his family. Why not out-scale this MO to include historical eras?

The Lost Tribes went to the Isles to prepare for the birth of Christianity. Those Druid priests "instantly" accepted the new system of belief brought to them by Joseph of Arimathea and the young Jesus. Isn't it a wonder that a national religion, centuries old would overnight melt into the English fog? Religions are harder to defeat than nations. The only explanation can be that Druidism was recognized as a corruption of God's Truth. It was so close to this "new" message that Druidism was immediately assimilated. One of the early converts to Christianity was the Arch Druid King Bran. He had stepped off the throne to become Arch Druid.

The Big Picture is that God sent Shem to prepare a fertile place for the birth of a nation (and then to another place to pre-establish the nation's capitol). In that place God built up the family of Abraham until they reached a size great enough to make a nation. Bringing them to a place of independence, He taught that nation and strengthened it until it was time to prepare the way for Jesus. He led His people to northwest Europe and the British Isles, where they could "renew your strength" (Isaiah). In the process, God dropped "seeds" in all the other places on earth; akin to remote receiving stations waiting for the word to come.

Once this network was in place, Jesus arrived to prove God's reality and tell His message of the coming earthly kingdom. The message was birthed in Jerusalem, grew up in England, and, via the Israelite world network, spread around the globe. And now with satellite broadcasting (done mainly by Israelites), this message of the coming kingdom has touched every square inch of the earth.

"And then shall the end come." Praise God!

"...stand still, and see the salvation of the Lord..."

It's clear that Jesus' coming, crucifixion and resurrection had many effects on the world and history. But what was his Mission? When he prayed to God in the garden that night and stated that he's accomplished God's directive, this was before his crucifixion. So we can throw out all his redemptive work. That hadn't happened yet. Well, what was Jesus' mission? If he didn't come to save us from our sins by dying, and he told God he'd done what he was sent to do, we should be able to find out what he did. It should be something big enough to include the whole world. We'll have to search Jesus' activities during that three and half years before he died. What did he come to do? Surely his activities will include acts that pertain to that Mission, even though there will be many other notable activities; like getting money out of a fish's mouth. What was Jesus mission?

Was it to cleanse the Temple? Hardly will this have worldwide ramifications.

Was it to feed the hungry? Lots of folks are still hungry.

Was it to heal? Then why didn't he heal more? And what about our sickness and infirmity?

Was it to cast out devils for the lucky few? (The word Luck comes out of Lucifer?)

Was it to show God's reality by doing miracles? Only God could raise the dead. Too many people are dead for this to be Jesus' Mission. Besides, proving God's reality wasn't necessary back then.

These last five activities of Christ, while of value, don't fit with his statement of the Mission. Was the cleansing, feeding, healing, raising, and casting done for "lost sheep of the House of Israel?"

Was it to go preaching the Kingdom of God; clearly one of his main occupations? Now this, finally, can expand into the world frame. God's Kingdom will be the Earth, not one Nation. And though there is ample scripture to show that Jesus went preaching the Kingdom of God, what happened to that message after his death?

Was it to train up a small band of disciples? Was the Mission to give them certain wisdom to begin disseminating to the rest of the world?

Was it to send these disciples to humanity with a message of worldwide significance? It must have been so. But what was that message? It wasn't the message of salvation. He hadn't been crucified yet, remember? The message had to be that of the coming Kingdom.

Was it to pass the Wand of Dissemination of God's Message from the Judahites to Britain? Britain? How did we get from God's coming Kingdom to Britain? Via those Apostles. They had no choice but to take that message to the "lost sheep of the House of Israel." Those were Jesus' explicit instructions. You'll find for the most part that the Apostles went to the places where the Lost Tribes of Israel had settled. But the great bulk of the Lost Tribes people would come to live in the British Isles and northwest Europe, before spreading out to become the British Empire.

And even though the Message of God's Kingdom has been relegated in Christianity to a position inferior to salvation, it is still God's number One priority. Setting up His earthly Kingdom is the original plan that God began with the first creation. That was when He created the Angels and the original Universe, the whole of which was marred by Lucifer's Fall. It's the place in Jeremiah that has cities but no man. (4:23-26) With the help of His angels, God was in the process of filling out and decorating the original Universe. The angels messed up and God gave the job to Adam. We know what happened to him.

Ever since, God has been coaxing humanity along a path to accomplish that same objective, enhancing creation. He's taking seven thousand years to do it. At the end of the Millennium, when the job of training folks to accomplish that work will be finished, God will start over with a new earth and universe for His helpers to work on.

God has been using Britain, northwest Europe and the USA to let the rest of the world know about His coming Kingdom. If Christ's Mission was to start the ball rolling in the House of Israel, then Britain, et al must be the House of Israel. But just stating the fact isn't quite satisfactory. It leaves out too much detail that will give us unflinching certainty that Christ was sent for those Lost Sheep.

So, we should be able to find a lot of Old Testament prophecy on the subject. All we have to do is nail down who are the people of that House of Israel. That should point us to much detail regarding that mission of Jesus.

If there's any scripture that is Messianic, it's the Book of Isaiah. So we should be able to find that House of Israel in Isaiah, along with details about the Messiah and His relationship to the House of Israel.

And there's no disappointment in the volume of information about the interaction of the Messiah and that House of Israel in the Book of Israel, rather Isaiah. Also some little nuggets, like the USA.

The book of Isaiah could easily be called the Book of Israel or the Book of Messiah.

Now let's go through the book of Israel and see if Jesus told the truth about being "not sent but for the lost sheep of the House of Israel." I'll only list about twenty-five of the references that connect Israel and Britain with the Messiah. But for clarification, I've included one passage from Jeremiah. We need this to keep us aware that God divorced the House of Israel, so that Isaiah 50 and 54 will make proper sense.

Amazingly, in the early chapters we'll be confronted with the United States!

We'll start out with Jeremiah 3:6-12. After clearly stating that there are two groups of people being talked about, "backsliding Israel" and "her treacherous sister Judah," the eighth verse says…"Israel committed adultery I had put her away, and given her a bill of divorce." If you want to see the terms of this divorce, read the first chapter of Hosea.

HISTORY IN BRIEF:

Israel sins big time. Assyria conquers them, re-locates them below the Black and Caspian seas. (Ho 1:4) They trek across Europe to the northwest and the British Isles (Ho 1:4 & 6), the last "official" arrival being the Normans in 1066 AD. (Ho 1:9) Joseph of Arimathea brings Christianity to Britain in 37 AD. This word is spread to the other Israelite nations occupying that northern and western region. They become the Christian nations of the world. (Ho 1:10) They bring the word of God's coming kingdom to the rest of the earth. Jesus comes a second time to re-marry and gather the Israelites back to the holy land to help him rule the earth for the thousand year Millennium. (Ho 1:11) Now Isaiah.

Isaiah 7 and 8 begin Israel's story. Isaiah is sent with his first and second sons to prophesy the destruction of the House of Israel as a nation. And please notice the enhancement of this prophecy encoded in Isaiah's son's names. Isaiah tells Ahaz that Israel will fall in 65 years. His son Shear-Jashub was with him. The name means "the remnant shall return", and signifies Israel's eventual return from their temporary home. Even though Ahaz won't be around to see it, God makes first mention of the birth of Christ in verse 14. In chapter eight Maher-shal'al-hash'-baz, whose name means "hasten the booty", accompanies Isaiah to confirm the prophesied destruction and spoiling of the northern kingdom of the House of Israel.

God gave all twelve tribes to Jesus. He calls them the "children thou hast given me." (John 5:36)

Isaiah 8:13-18 Jesus, the "rock of offense" (v14) will "wait upon the LORD (God)." (v17) Jesus and the "hidden" House of Israel will be signs from God. In other words, God is going to send Immanuel to divorced and hidden Israel.

Isaiah 9:8-9 This passage follows up the prophecy of the Messiah, telling that God sends Jesus, the Light, and he's known by Israel, the ten tribes. Israel returns to God, converts to Christianity.

Isaiah 11:10-16 After the Messiah comes, (v10) Israel and Judah, in peace, are gathered back to Palestine to rule all their old enemies, under Christ. This passage again points up the fact of two distinct groups of people, the Israelites and the Jews/Judahites.

Isaiah 14:1-3 This section tells what will happen to the House of Israel in their home away from Home.

Israel will get:
    Mercy
    Chosen
    Their own land
    Strangers will serve them in that land

> They will take captive their captors
> They will rule their oppressors
> And have rest from sorrow, fear, and bondage.

We must ask, "Do all these promises fit the Jews of history?"
> OK
> No-they were never un-chosen
> No-not 'till 1948
> No-not 'till after 1948
> No-only in modern Israel
> No-only in modern Israel
> No-the Jews have been on the edge of their seats since they became a nation in 1948

Isaiah 18:1-7 This is the section that tells of the USA so we'll have to go slow here. We have five attributes listed:
> Land of the eagle.
> Sends ambassadors by ships of bulrushes
> Scattered and peeled
> Terrible form their beginning
> Meted out and trodden down
> The rivers have spoiled.
> The world sees it's ensign
> The world hears it's trumpet

Analysis:

> Britain, the Center of the House of Israel has an Ox or a Unicorn, not an eagle. The US has an Eagle as its emblem.
> Bulrushes suck up water. What kind of ship uses "water" power? A steamship.
> The marginal reading for scattered and peeled is "out-spread and polished." The US is a large, beautiful land.
> The US began as a country by the use of terrible force. That's not the way England came into existence. Technically speaking, Britain has been several different countries over the centuries.
> Two meanings have been given to "meted out and trodden down." Surveyed and apportioned, and stalwart and strong. Strong's Concordance will help you sort this one out.
> The word spoiled is the same liturgical word used for quartering the sacrifice. The US is virtually quartered by its rivers. The

Mississippi, Platt/Colorado and Ohio come pretty close.
When the US has raised its banner, the world has always taken notice.
Likewise, when the US blows its trumpet of war, the world pays attention.

What other Israelite nation would you nominate? Britain only fits some of these characteristics, and it's the only other Israelite nation that comes close. Certainly Holland or Denmark aren't large, terrible in war or have an eagle as their emblem. As hard as it is to imagine, Isaiah prophesies about a major portion of the Lost Tribes of the House of Israel, the United States of America. Lastly, does this description fit Judah? No!

Isaiah 37:31 Here God sneaks in another reference to Judah, telling how the kingly lines of Zarah and Pharez will join in the development of the House of Israel. "A remnant of Judah shall take root down, and bear fruit up." This refers to Jeremiah's mission to Ireland to plant the Pharez-Judah daughter of Zedekiah, in marriage, with the Zarah-Judah-Milesian Eochaide. This is followed up in chapter 41.

Isaiah 41:1-9 The Isles, in their new land, will "renew their strength." Then verse two begins an enigmatic passage, which has been mistakenly ascribed to Cyrus, who accomplished the release of the House of Judah from their captivity in Babylon. This section describes Jeremiah and his trip to Ireland. Cyrus may have been God's chosen instrument for the release of the Jews, but he isn't given the name "righteous." Jeremiah was to tell kings what to do, in Israel and in Ireland. And finally, Jeremiah had to get to his destination by ship, not foot. Cyrus didn't have to sail over to Babylon to conquer it. The passage ends with the repeated promise of God to bring Israel home in the end and not cast them away completely.

Now we begin more intensely to see the relationship of Christ to the House of Israel. From here to the end of the Book we see the interaction between the Messiah and Israel.

Isaiah 42:1--> The first four verses talk clearly of Christ. After four and three quarters verses, it says that, "the isles shall wait for his law." The British Isles will renew their strength while they wait for the first advent of Jesus. The isles will sing. (v10) The islands will declare God's praise. (v12) Verse sixteen describes the migrations of the Lost Tribes.

Isaiah 44:1-6 God pours out His Spirit on Israel. Don't be confused by all the different names given to Israel. For confirmation, see Joel 2:28-29, which

Luke quotes in Acts 2:17-18 and Jeremiah alludes to in 31:33 when he tells of the "new covenant" that God will make with Israel.

Isaiah 49:1-6 The chapter starts out talking to "the isles," and clearly names them in verse three, "Israel." Then He talks about what Israel will accomplish. (v6)
    Raise up the tribes
    Restore the preserved of Israel
    Be a light to the Gentiles
    Be God's salvation to the end of the earth.

Isaiah 49:7-12 Christ is given to Israel and they come back form "the north and west." (v12) This is right where the British Isles are located.

Isaiah 49:19-20 This passage reiterates how Israel's former enemies will be far away, but includes another interesting bit of future-telling. Israel is promised to be a "company of nations." This turns out to be the British Empire. But we know that this mightiest and most extensive Empire of history eventually began to break up. It all started with that nation "terrible from its beginning." After the US revolted, other members of the Empire sued for their independence also. That's what Isaiah foretells in verse 20. "the children (nations) which thou shalt have, after thou hast lost the other, shall say again in thine ears, 'The place is too strait for me; give place to me that I may dwell."

Isaiah 50:1 Can there be any doubt that this is the same Israel that was divorced by God in Jeremiah?

Isaiah 51:1-6 In the midst of informing all His people, God singles out the isles again to say that they will wait upon His law, which has previously been shown to be Christ. This law is to be a "light of the people."

Isaiah 52:--> God spends much time announcing a lot of good stuff for Zion and Jerusalem. Then He names His "servant" who will accomplish this. (v13) Yes Jesus will come to Jerusalem and will initiate all those good things, but what will happen to him?

Isaiah 53:--> We are given a very detailed account of what happens to the Messiah. So detailed that this chapter can't be mistaken. This is the famous Suffering Servant chapter, the one Handel used in his Messiah. Now God can sidestep His bill of divorce that He gave Israel; His own law states that a divorced person cannot remarry until the death of the other spouse. When

Jesus, as God, died, the door to divorced Israel was opened. Now God can remarry Israel. That's why the 54th chapter starts out "Sing oh ye barren."

Isaiah 54:1-3 the Barren, divorced wife Israel has actually had a multitude of children. Many more than Judah, who remained married to God.

Isaiah 54:5-6 Israel's Maker is now her Husband; a wife of youth that was forsaken.

Isaiah 54:13-17 Here we have yet another list by which to judge the real Israel.
    Israel's kids will taught of God.
    They will be at peace.
    Shall be righteous.
    Shall be far from oppression.
    Shall not fear.
    Their enemies shall fall.
    Never be totally conquered "no weapon formed...shall prevail." (Jer 51:20)
    Shall condemn those who judge her.

I especially like the "no weapon" part. Who put down world tyranny from the time of Christ to the present? Who was responsible for the fall of Rome, Napoleon, the Kaiser, Hitler, Tojo? Britain, the US and their allies. Israel has been "God's battle axe." (Jer 51:20)

Isaiah 62:2-5 Verse four tells who this passage is talking about. It's about the "Forsaken." (v4) Israel shall be called by a new name (not Israel anymore). One name will be Hephzibah, which means "My delight is in her." Also Israel shall be called Beulah, Married. The House of Israel is the only divorced wife, and can be the only one to get married. Lastly, in verse five, we have the term often associated with Christ, "the bridegroom."

There are many more verses that can fill in this story of Jesus and his mission to the House of Israel. To me, the above is conclusive evidence of the truth of Jesus' statement about his mission to the earth.

He said, "I am not sent but for the lost sheep of the House of Israel." (Matt 15:24) He told his disciples, "But go rather to the lost sheep of the house of Israel." Why was that?

God was still accomplishing His plan for setting up a kingdom here on earth.

He still needed to get that information out to the rest of the world. The Jews were clearly not taking that message forward, so it was up to the House of Israel.

Jesus went preaching, not salvation, but the kingdom of God. Check all the references to Jesus' preaching. His main job was to get the ball rolling so that this message could get out to the rest of the world. God had tucked away the House of Israel for just that task. Looking through the back door, we find that the two groups of people in all of history that disseminated the news of God and His coming kingdom were the British and the Americans. The House of Israel.

The Book Of Isaiah is indeed the Book of Israel.

The famed Roman-Jewish historian, Flavius Josephus, said in the first century AD:

"Wherefore there are but two [Israel] tribes in Europe and Asia subject to the Romans, while the ten tribes are beyond the Euphrates till now, and are ***an immense multitude, and not to be estimated by numbers***." (10)

Famed early 20th century historian archaeologist, Archibald Henry Sayce, in his book, *Higher Critics and the Monuments*, p. 396, adds:

"It was, however, in the time of Ahab the son of Omri that the Assyrians first became acquainted with the kingdom of Israel, and consequently Samaria continued ever afterwards to be known to them as Beth-Omri, the 'house of Omri' (or, 'Khumri')."

"... the Sacae, or Scythians, who, again, were the Lost Ten Tribes."
The Jewish Encyclopaedia, Vol 12. p250

Israel = Saka = Gimirri = Khumri

Note also what *The Jewish Encyclopaedia* Vol. 12, p.250 states:

"The identification of the Sacae, or Scythians with the Ten Tribes because they appear in history at the same time, and very nearly in the same place as the Israelites removed by Shalmaneser, is one of the chief-supports of the theory which identifies the English-people, and indeed the whole Teutonic race, with the Ten Tribes."

Proofs abound as to where are these 'missing tribes'. The apocryphal book of First Maccabees ch. 12 v.21 (from about 100BC) states that:

"It has been found in writing concerning the Spartans [Greeks] and the Jews (Judahites!)that they are brethren, and that they are of the stock of Abraham." Similar connections abound between ancient Israel and Italy, Spain, Ireland, Britain, and other countries in Europe.

If this is true, then should the Anglo-Saxons have Jewish features? On this, the *Westminster Historical Atlas To The Bible* reproduces an ancient inscription from the Temple of Rameses III at Medinet Habu in Egypt, which states;

"Canaanite captives in Egypt being led before the Pharaoh. The relief, which portrays the general appearance of Israelites as well as Canaanites, is a good representation of the typical Semite of the day. Note the noble, aristocratic features, particularly the finely cut noses and the long hair and beards. It is commonly thought that Israelites had 'hooked noses', but this was originally a Hittite or Armenoid feature." (13)

THE ENTIRE LAW OF YHWH (GOD)

These "statutes" are separate from the Ten Commandments known as THE LAW. These Statutes were given to the Israelites through MOSES and the high priests as received from Yahweh God. Closer observation shows that much of the life in Egypt over 400 years led the Israelites astray with regard to idol worship, diets and a host of other habits formed that had to be broken and replaced. God was establishing Himself once again into the people of Israel and was to make them a great and mighty nation again!

GOD

1. To know that God exists (Ex. 20:2; Deut. 5:6)
2. Not to entertain the idea that there is any god but the Eternal (Ex. 20:3)
3. Not to blaspheme (Ex. 22:27-28) Cf Lev 24:16
4. To hallow God's name (Lev. 22:32)
5. Not to profane God's name (Lev. 22:32
6. To know that God is One, a complete Unity (Deut. 6:4)
7. To love God (Deut. 6:5)
8. To fear Him reverently (Deut. 6:13; 10:20)
9. Not to put the word of God to the test (Deut. 6:16)
10. To imitate His good and upright ways (Deut. 28:9)

The Law

11. To honor the old and the wise (Lev. 19:32)
12. To learn Torah and to teach it (Deut. 6:7)

13. To cleave to those who know Him (Deut. 10:20)
14. Not to add to the commandments of the Torah.(Deut.13:1)
15. Not to take away from the commandments of the Torah (Deut.13:1)
16. That every person shall write a scroll of the Torah for himself (Deut. 31:19)

Signs and Symbols

17. To circumcise the male offspring (Gen. 17:12; Lev. 12:3)
18. To put fringes on the corners of clothing (Num. 15:38)
19. To bind God's Word on the head (Deut. 6:8)
20. To bind God's Word on the arm (Deut. 6:8)
21. To affix the mezuzah to the door posts and gates of your house (Deut. 6:9)

Prayer and Blessing

22. To pray to God (Ex. 23:25; Deut. 6:13)
23. To read the *Shema* [lit: The Hearing] in the morning and at night (Deut. 6:7)
24. To recite grace after meals (Deut. 8:10)
25. Not to lay down a stone for worship (Lev. 26:1)

Love and Brotherhood

26. To love all human beings who are of the covenant (Lev. 19:18)
27. Not to stand by idly when a human life is in danger (Lev. 19:16)
28. Not to wrong any one in speech (Lev. 25:17)
29. Not to carry tales (Lev. 19:16)
30. Not to cherish hatred in one's heart (Lev. 19:17)
31. Not to take revenge (Lev. 19:18)
32. Not to bear a grudge (Lev. 19:18)
33. Not to put any Israelite to shame (Lev. 19:17)
34. Not to curse any other Israelite (Lev. 19:14)
35. Not to give occasion to the simple-minded to stumble on the road (Lev. 19:14) (this includes doing anything that will cause another to sin)
36. To rebuke the sinner (Lev. 19:17)
37. To relieve a neighbor of his burden and help to unload his beast (Ex. 23:5)
38. To assist in replacing the load upon a neighbor's beast (Deut. 22:4)
39. Not to leave a beast, that has fallen down beneath its burden, unaided (Deut. 22:4)

The Poor and Unfortunate

40. Not to afflict an orphan or a widow (Ex. 22:21)
41. Not to reap the entire field (Lev. 19:9; Lev. 23:22)
42. To leave the unreaped corners of the field or orchard for the poor (Lev. 19:9)
43. Not to gather gleanings (the ears that have fallen

to the ground while reaping)
(Lev. 19:9)
44. To leave the gleanings for the poor (Lev. 19:9)
45. Not to gather ol'loth (the imperfect clusters) of the vineyard (Lev. 19:10)
46. To leave ol'loth (the imperfect clusters) of the vineyard for the poor (Lev. 19:10; Deut. 24:21)
47. Not to gather the single grapes that have fallen to the ground (Lev. 19:10)
48. To leave the single grapes of the vineyard for the poor (Lev. 19:10)
49. Not to return to take a forgotten sheaf (Deut. 24:19) This applies to all fruit trees (Deut. 24:20)
50. To leave the forgotten sheaves for the poor (Deut. 24:19-20)
51. Not to refrain from maintaining a poor man and giving him what he needs (Deut. 15:7)
52. To give charity according to one's means (Deut. 15:11)

Treatment of the Non-Israelites

53. To love the stranger (Deut. 10:19) (CCA61).
54. Not to wrong the stranger in speech (Ex. 22:20)
55. Not to wrong the stranger in buying or selling (Ex. 22:20)
56. Not to intermarry with non-Israelites (Deut. 7:3)
57. To exact the debt of an alien (Deut. 15:3)
58. To lend to an alien at interest (Deut. 23:21)

Marriage, Divorce and Family

59. To honor father and mother (Ex. 20:12)
60. Not to smite a father or a mother (Ex. 21:15)
61. Not to curse a father or mother (Ex. 21:17)
62. To reverently fear father and mother (Lev. 19:3)
63. To be fruitful and multiply (Gen. 1:28)
64. That a eunuch shall not marry a daughter of Israel (Deut. 23:2)
65. That a bastard [Heb.mamzer = illegitimate son] shall not marry the daughter of a Israelite (Deut.23:3)
66. That an Ammonite or Moabite shall never marry the daughter of an Israelite (Deut. 23:4)
67. Not to exclude a descendant of Esau from the community of Israel for three generations (Deut. 23:8-9)
68. Not to exclude an Egyptian from the community of Israel for three generations (Deut. 23:8-9)
69. That there shall be no harlot (in Israel); that is, that there shall be no intercourse with a woman, without previous marriage with a deed of marriage and formal declaration of marriage (Deut.23:18)
70. To take a wife by the sacrament of marriage (Deut.24:1)
71. That the newly married husband shall (be free) for one year to rejoice with his wife (Deut. 24:5)
72. That a bridegroom shall be exempt for a whole year from taking part in any public labor, such as military service, guarding the wall and similar duties (Deut.

24:5)
73. Not to withhold food, clothing or conjugal rights from a wife (Ex. 21:10)
74. That the woman suspected of adultery shall be dealt with as prescribed in the Torah (Num. 5:30)
75. That one who defames his wife's honor (by falsely accusing her of unchastity before marriage) must live with her all his lifetime (Deut. 22:19)
76. That a man may not divorce his wife concerning whom he has published an evil report (about her unchastity before marriage) (Deut. 22:19)
77. To divorce by a formal written document (Deut. 24:1)
78. That one who divorced his wife shall not remarry her, if after the divorce she had been married to another man (Deut. 24:4)
79. That a widow whose husband died childless must not be married to anyone but her deceased husband's brother (Deut. 25:5) (this is only in effect insofar as it requires the procedure of release below).
80. To marry the widow of a brother who has died childless (Deut.25:5)
(this is only in effect insofar as it requires the procedure of release below)
81. That the widow formally release the brother-in-law (if he refuses to marry her) (Deut. 25:7-9)

Forbidden Sexual Relations

82. Not to indulge in familiarities with relatives, such

as sensual kissing, carnal embracing, or provocative winking which may lead to incest (Lev.18:6)
83. Not to commit incest with one's mother (Lev. 18:7)
84. Not to commit sodomy with one's father (Lev. 18:7)
85. Not to commit incest with one's father's wife (Lev. 18:8)
86. Not to commit incest with one's sister (Lev. 18:9)
87. Not to commit incest with one's father's wife's daughter (Lev.18:9)
88. Not to commit incest with one's son's daughter (Lev. 18:10)
89. Not to commit incest with one's daughter's daughter (Lev.18:10)
90. Not to commit incest with one's daughter (this is not explicitly in the Torah but is inferred from other explicit commands that would include it)
91. Not to commit incest with one's fathers sister (Lev. 18:12)
92. Not to commit incest with one's mother's sister (Lev. 18:13)
93. Not to commit incest with one's father's brothers wife (Lev.18:14)
94. Not to commit sodomy with one's father's brother (Lev. 18:14)
95. Not to commit incest with one's son's wife (Lev. 18:15)
96. Not to commit incest with one's brother's wife (Lev. 18:16)
97. Not to commit incest with one's wife's daughter

(Lev. 18:17)
98. Not to commit incest with the daughter of one's wife's son (Lev.18:17)
99. Not to commit incest with the daughter of one's wife's daughter (Lev. 18:17)
100. Not to commit incest with one's wife's sister (Lev. 18:18)
101. Not to have intercourse with a woman, in her menstrual period (Lev. 18:19)
102. Not to have intercourse with another man's wife (Lev. 18:20)
103. Not to commit sodomy with a male (Lev. 18:22)
104. Not to have intercourse with a beast (Lev. 18:23)
105. That a woman shall not have intercourse with a beast (Lev.18:23)
106. Not to castrate the male of any species; neither a man, nor a domestic or wild beast, nor a fowl (Lev. 22:24)

Times and Seasons

107. That the new month shall be solemnly proclaimed as holy, and the months and years shall be calculated by the Supreme Court only (Ex. 12:2)
108. Not to travel on the Sabbath outside the limits of one's place of residence
(Ex. 16:29)
109. To sanctify the Sabbath (Ex. 20:8)
110. Not to do work on Sabbath (Ex. 20:10)

111. To rest on Sabbath (Ex. 23:12; 34:21)
112. To celebrate the festivals (Ex.23:14)
113. To rejoice on the festivals (Deut. 16:14)
114. To appear in the Sanctuary on the festivals (Deut. 16:16)
115. To remove leaven on the Eve of Passover (Ex. 12:15)
116. To rest on the first day of Passover (Ex. 12:16; Lev. 23:7)
117. Not to do work on the first day of Passover (Ex. 12:16; Lev.23:6-7)
118. To rest on the seventh day of Passover (Ex. 12:16; Lev. 23:8)
119. Not to do work on the seventh day of Passover (Ex. 12:16;Lev. 23:8)
120. To eat "matzah" [unleavened bread] on the first night of Passover (Ex. 12:18)
121. That no leaven be in the Israelite's possession during Passover (Ex. 12:19)
122. Not to eat any food containing leaven on Passover (Ex.12:20)
123. Not to eat leaven on Passover (Ex. 13:3)
124. That leaven shall not be seen in an Israelite's home during Passover (Ex. 13:7)
125. To discuss the departure from Egypt on the first night of Passover (Ex. 13:8)
126. Not to eat leaven after mid-day on the fourteenth of Nissan (Deut. 16:3)
127. To count forty-nine days from the time of the cutting of the Omer

(i.e. first sheaves of the barley harvest) (Lev. 23:15)
128. To rest on Pentecost (Lev. 23:21)
129. Not to do work on the feast of Pentecost (Lev. 23:21)
130. To rest on Rosh Hashanah [i.e the feast of Trumpets] (Lev. 23:24) (CCA29)
131. Not to do work on Rosh Hashanah (Lev. 23:25)
132. To hear the sound of the Trumpet [Heb. shofar or ram's horn] (Num.29:1)
133. To fast on Yom Kippur i.e the day of Atonement (Lev. 23:27)
134. Not to eat or drink on Yom Kippur (Lev. 23:29) (CCN152)
135. Not to do work on Yom Kippur (Lev. 23:31) (CCN151)
136. To rest on the Yom Kippur (Lev. 23:32)
137. To rest on the first day of the feast of Tabernacles or Booths.[Heb. Sukkot] (Lev. 23:35)
138. Not to do work on the first day of the feast of Tabernacles. (Lev. 23:35)
139. To rest on the eighth day of the feast of Tabernacles (Lev.23:36)
140. Not to do work on the eighth day of the feast of Tabernacles
(Lev. 23:36)
141. To take during Sukkot a palm branch and the other three plants (Lev. 23:40)
142. To dwell in booths seven days during Sukkot (Lev. 23:42

Dietary Laws

143. To examine the marks in cattle (so as to distinguish the clean from the unclean) (Lev. 11:2)
144. Not to eat the flesh of unclean beasts (Lev. 11:4)
145. To examine the marks in fishes (so as to distinguish the clean from the unclean (Lev. 11:9)
146. Not to eat unclean fish (Lev. 11:11)
147. To examine the marks in fowl, so as to distinguish the clean from the unclean (Deut. 14:11)
148. Not to eat unclean fowl (Lev. 11:13)
149. To examine the marks in locusts, so as to distinguish the clean from the unclean (Lev. 11:21)
150. Not to eat a worm found in fruit (Lev. 11:41)
151. Not to eat of things that creep upon the earth (Lev. 11:41-42)
152. Not to eat any vermin of the earth (Lev. 11:44)
153. Not to eat things that swarm in the water (Lev. 11:43 and 46)
154. Not to eat of winged insects (Deut. 14:19)
155. Not to eat the flesh of a beast that is torn (Ex. 22:30)
156. Not to eat the flesh of a beast that died of itself (Deut. 14:21)
157. To slay cattle, deer and fowl according to the law if their flesh is to be eaten (Deut. 12:21)
158. Not to eat a limb removed from a living beast (Deut. 12:23)
159. Not to slaughter an animal and its young on the

same day (Lev.22:28)
160. Not to take the mother-bird with the young (Deut. 22:6)
161. To set the mother-bird free when taking the nest (Deut.22:6-7)
162. Not to eat the flesh of an ox that was condemned to be stoned (Ex. 21:28)
163. Not to boil meat with milk (Ex. 23:19)
164. Not to eat flesh with milk (Ex. 34:26)
165. Not to eat the of the thigh-vein which shrank (Gen. 32:33)
166. Not to eat the fat of the offering (Lev. 7:23)
167. Not to eat blood (Lev. 7:26)
168. To cover the blood of undomesticated animals (deer, etc.) and of fowl that have been killed (Lev. 17:13)
169. Not to eat or drink like a glutton or a drunkard (not to rebel against father or mother) (Lev. 19:26; Deut. 21:20)

Business Practices

170. Not to do wrong in buying or selling (Lev. 25:14)
171. Not to make a loan to an Israelite on interest (Lev. 25:37)
172. Not to borrow on interest (Deut. 23:20) (because this would cause the lender to sin)
173. Not to take part in any usurious transaction between borrower and lender, neither as a surety, nor

as a witness, nor as a writer of the bond for them (Ex. 22:24)
174. To lend to a poor person (Ex. 22:24)
175. Not to demand from a poor man repayment of his debt, when the creditor knows that he cannot pay, nor press him (Ex.22:24)
176. Not to take in pledge utensils used in preparing food (Deut.24:6)
177. Not to exact a pledge from a debtor by force (Deut. 24:10)
178. Not to keep the pledge from its owner at the time when he needs it (Deut. 24:12)
179. To return a pledge to its owner (Deut. 24:13)
180. Not to take a pledge from a widow (Deut. 24:17)
181. Not to commit fraud in measuring (Lev. 19:35)
182. To ensure that scales and weights are correct (Lev. 19:36)
183. Not to possess inaccurate measures and weights (Deut.25:13-14)

Employees, Servants and Slaves

184. Not to delay payment of a hired man's wages (Lev. 19:13)
185. That the hired laborer shall be permitted to eat of the produce he is reaping (Deut. 23:25-26)
186. That the hired laborer shall not take more than he can eat (Deut. 23:25)
187. That a hired laborer shall not eat produce that is

not being harvested (Deut. 23:26)
188. To pay wages to the hired man at the due time (Deut. 24:15)
189. To deal judicially with the Hebrew bondman in accordance with the laws appertaining to him (Ex. 21:2-6)
190. Not to compel the Hebrew servant to do the work of a slave (Lev. 25:39)
191. Not to sell a Hebrew servant as a slave (Lev. 25:42)
192. Not to treat a Hebrew servant rigorously (Lev. 25:43)
193. Not to permit a gentile to treat harshly a Hebrew bondman sold to him (Lev. 25:53)
194. Not to send away a Hebrew bondman servant empty handed, when he is freed from service (Deut. 15:13)
195. To bestow liberal gifts upon the Hebrew bondsman (at the end of his term of service), and the same should be done to a Hebrew bondwoman (Deut. 15:14)
196. To redeem a Hebrew maid-servant (Ex. 21:8)
197. Not to sell a Hebrew maid-servant to another person (Ex. 21:8)
198. To espouse a Hebrew maid-servant (Ex. 21:8-9)
199. To keep the Canaanite slave forever (Lev. 25:46)
200. Not to surrender a slave, who has fled to the land of Israel, to his owner who lives outside Palestine (Deut. 23:16)
201. Not to wrong such a slave (Deut. 23:17)
202. Not to muzzle a beast, while it is working in

produce which it can eat and enjoy (Deut. 25:4)

Vows, Oaths and Swearing

203. That a man should fulfill whatever he has uttered (Deut. 23:24)
204. Not to swear needlessly (Ex. 20:7)
205. Not to violate an oath or swear falsely (Lev. 19:12)
206. To decide in cases of annulment of vows, according to the rules set forth in the Torah (Num. 30:2-17)
207. Not to break a vow (Num. 30:3)
208. To swear by His name truly (Deut. 10:20)
209. Not to delay in fulfilling vows or bringing vowed or free-will offerings (Deut. 23:22)

The Sabbatical and Jubilee Years

210. To let the land lie fallow in the Sabbatical year (Ex. 23:11; Lev.25:2)
211. To cease from tilling the land in the Sabbatical year (Ex. 23:11) (Lev. 25:2)
212. Not to till the ground in the Sabbatical year (Lev. 25:4)
213. Not to do any work on the trees in the Sabbatical year (Lev.25:4)
214. Not to reap the aftermath that grows in the Sabbatical year, in the same way as it is reaped in other years (Lev. 25:5)
215. Not to gather the fruit of the tree in the

Sabbatical year in the same way as it is gathered in other years (Lev. 25:5)
216. To sound the Ram's horn in the Sabbatical year (Lev. 25:9)
217. To release debts in the seventh year (Deut. 15:2)
218. Not to demand return of a loan after the Sabbatical year has passed (Deut. 15:2)
219. Not to refrain from making a loan to a poor man, because of the release of loans in the Sabbatical year (Deut. 15:9)
220. To assemble the people to hear the Torah at the close of the seventh year (Deut. 31:12)
221. To count the years of the Jubilee by years and by cycles of seven years (Lev. 25:8)
222. To keep the Jubilee year holy by resting and letting the land lie fallow (Lev. 25:10)
223. Not to cultivate the soil nor do any work on the trees, in the Jubilee Year (Lev. 25:11)
224. Not to reap the aftermath of the field that grew of itself in the Jubilee Year, in the same way as in other years (Lev. 25:11)
225. Not to gather the fruit of the tree in the Jubilee Year, in the same way as in other years (Lev. 25:11)
226. To grant redemption to the land in the Jubilee year (Lev. 25:24)

The Court and Judicial Procedure

227. To appoint judges and officers in every community

of Israel (Deut. 16:18)
228. Not to appoint as a judge, a person who is not well versed in the laws of the Torah, even if he is expert in other branches of knowledge (Deut. 1:17)
229. To adjudicate cases of purchase and sale (Lev. 25:14)
230. To judge cases of liability of a paid depositary (Ex. 22:9)
231. To adjudicate cases of loss for which a gratuitous borrower is liable (Ex. 22:13-14)
232. To adjudicate cases of inheritances (Num. 27:8-11)
233. To judge cases of damage caused by an uncovered pit (Ex.21:33-34)
234. To judge cases of injuries caused by beasts (Ex. 21:35-36)
235. To adjudicate cases of damage caused by trespass of cattle (Ex.22:4)
236. To adjudicate cases of damage caused by fire (Ex. 22:5)
237. To adjudicate cases of damage caused by a gratuitous depositary (Ex. 22:6-7)
238. To adjudicate other cases between a plaintiff and a defendant (Ex. 22:8)
239. Not to curse a judge (Ex. 22:27)
240. That one who possesses evidence shall testify in Court (Lev.5:1)
241. Not to testify falsely (Ex. 20:13)
242. That a witness, who has testified in a capital case, shall not lay down the law in that particular case (Num.

35:30)
243. That a transgressor shall not testify (Ex. 23:1)
244. That the court shall not accept the testimony of a close relative of the defendant in matters of capital punishment (Deut. 24:16)
245. Not to hear one of the parties to a suit in the absence of the other party (Ex. 23:1)
246. To examine witnesses thoroughly (Deut. 13:15)
247. Not to decide a case on the evidence of a single witness (Deut.19:15)
248. To give the decision according to the majority, when there is a difference of opinion among the members of the Sanhedrin as to matters of law (Ex. 23:2)
249. Not to decide, in capital cases, according to the view of the majority, when those who are for condemnation exceed by one only, those who are for acquittal (Ex. 23:2)
250. That, in capital cases, one who had argued for acquittal, shall not later on argue for condemnation (Ex. 23:2)
251. To treat parties in a litigation with equal impartiality (Lev. 19:15)
252. Not to render iniquitous decisions (Lev. 19:15)
253. Not to favor a great man when trying a case (Lev. 19:15)
254. Not to take a bribe (Ex. 23:8)
255. Not to be afraid of a bad man, when trying a case (Deut. 1:17)

256. Not to be moved in trying a case, by the poverty of one of the parties (Ex. 23:3; Lev. 19:15)
257. Not to pervert the judgment of strangers or orphans (Deut. 24:17)
258. Not to pervert the judgment of a sinner (a person poor in fulfillment of commandments) (Ex. 23:6)
259. Not to render a decision on one's personal opinion, but only on the evidence of two witnesses, who saw what actually occurred (Ex. 23:7)
260. Not to execute one guilty of a capital offense, before he has stood his trial (Num. 35:12)
261. To accept the rulings of every Supreme Court in Israel (Deut. 17:11)
262. Not to rebel against the orders of the Court (Deut. 17:11)

Injuries and Damages

263. To make a parapet for your roof (Deut. 22:8)
264. Not to leave something that might cause hurt (Deut. 22:8)
265. To save the pursued even at the cost of the life of the pursuer (Deut. 25:12)
266. Not to spare a pursuer, but he is to be slain before he reaches the pursued and slays the latter, or uncovers his nakedness (Deut. 25:12)

Property and Property Rights

267. Not to sell a field in the land of Israel in perpetuity (Lev. 25:23)
268. Not to change the character of the open land (about the cities of) the Levites or of their fields; not to sell it in perpetuity, but it may be redeemed at any time (Lev. 25:34)
269. That houses sold within a walled city may be redeemed within a year (Lev. 25:29)
270. Not to remove landmarks (property boundaries) (Deut. 19:14)
271. Not to swear falsely in denial of another's property rights (Lev. 19:11)
272. Not to deny falsely another's property rights (Lev. 19:11)
273. Never to settle in the land of Egypt (Deut. 17:16)
274. Not to steal personal property (Lev. 19:11)
275. To restore that which one took by robbery (Lev. 5:23)
276. To return lost property (Deut. 22:1)
277. Not to pretend not to have seen lost property, to avoid the obligation to return it (Deut. 22:3)

Criminal Laws

278. Not to slay an innocent person (Ex. 20:13)
279. Not to kidnap any person of Israel (Ex. 20:13)
280. Not to rob by violence (Lev. 19:13)

281. Not to defraud (Lev. 19:13)
282. Not to covet what belongs to another (Ex. 20:14)
283. Not to crave something that belongs to another (Deut. 5:18)
284. Not to indulge in evil thoughts and sights (Num. 15:39)

Punishment and Restitution

285. That the Court shall pass sentence of death by decapitation with the sword (Ex. 21:20; Lev. 26:25)
286. That the Court shall pass sentence of death by strangulation (Lev. 20:10)
287. That the Court shall pass sentence of death by burning with fire (Lev. 20:14)
288. That the Court shall pass sentence of death by stoning (Deut. 22:24)
289. To hang the dead body of one who has incurred that penalty (Deut. 21:22)
290. That the dead body of an executed criminal shall not remain hanging on the tree overnight (Deut. 21:23)
291. To inter the executed on the day of execution (Deut. 21:23)
292. Not to accept ransom from a murderer (Num. 35:31)
293. To exile one who committed accidental homicide (Num. 35:25)
294. To establish six cities of refuge (for those who committed accidental homicide)

(Deut. 19:3)
295. Not to accept ransom from an accidental homicide, so as to relieve him from exile (Num. 35:32)
296. To decapitate the heifer in the manner prescribed (in expiation of a murder on the road, the perpetrator of which remained undiscovered) (Deut. 21:4)
297. Not to plow nor sow the rough valley (in which a heifer's neck was broken)
(Deut. 21:4)
298. To adjudge a thief to pay compensation or (in certain cases) suffer death (Ex. 21:16; Ex. 21:37; Ex. 22:1)
299. That he who inflicts a bodily injury shall pay monetary compensation (Ex. 21:18-19)
300. To impose a penalty of fifty shekels upon the seducer (of an unbetrothed virgin) and enforce the other rules in connection with the case (Ex. 22:15-16)
301. That the violator (of an unbetrothed virgin) shall marry her (Deut. 22:28-29)
302. That one who has raped a damsel and has then (in accordance with the law) married her, may not divorce her (Deut. 22:29)
303. Not to inflict punishment on the sabbath (Ex. 35:3) (because some punishments were inflicted by fire)
304. To punish the wicked by the infliction of stripes (Deut. 25:2)
305. Not to exceed the statutory number of stripes laid on one who has incurred that punishment (Deut.

25:3) (and by implication, not to strike anyone)
306. Not to spare the offender, in imposing the prescribed penalties on one who has caused damage (Deut. 19:13)
307. To do unto false witnesses as they had purposed to do (to the accused) (Deut. 19:19)
308. Not to punish any one who has committed an offense under duress (Deut. 22:26)

Prophecy

309. To heed the call of every prophet in each generation, provided that he neither adds to, nor takes away from the Torah (Deut. 18:15)
310. Not to prophesy falsely (Deut. 18:20)
311. Not to refrain from putting a false prophet to death nor to be in fear of him (Deut. 18:22) (negative)

Idolatry, Idolaters and Idolatrous Practices

312. Not to make a graven image; neither to make it oneself nor to have it made by others (Ex. 20:4)
313. Not to make any figures for worship, even if they are not worshipped (Ex. 20:20)
314. Not to make idols even for others (Ex. 34:17; Lev. 19:4)
315. Not to use the ornament of any object of idolatrous worship (Deut. 7:25)
316. Not to make use of an idol or its accessory objects, offerings, or libations (Deut. 7:26)
317. Not to drink wine of idolaters (Deut. 32:38)

318. Not to worship an idol in the way in which it is usually worshipped (Ex. 20:5)
319. Not to bow down to an idol, even if that is not its mode of worship (Ex. 20:5)
320. Not to prophesy in the name of an idol (Ex. 23:13; Deut.18:20)
321. Not to hearken to one who prophesies in the name of an idol (Deut. 13:4)
322. Not to lead the children of Israel astray to idolatry (Ex. 23:13)
323. Not to entice an Israelite to idolatry (Deut. 13:12)
324. To destroy idolatry and its appurtenances (Deut. 12:2-3)
325. Not to love the enticer to idolatry (Deut. 13:9)
326. Not to give up hating the enticer to idolatry (Deut. 13:9)
327. Not to save the enticer from capital punishment, but to stand by at his execution (Deut. 13:9)
328. A person whom he attempted to entice to idolatry shall not urge pleas for the acquittal of the enticer (Deut. 13:9)
329. A person whom he attempted to entice shall not refrain from giving evidence of the enticer's guilt, if he has such evidence (Deut. 13:9)
330. Not to swear by an idol to its worshipers, nor cause them to swear by it (Ex. 23:13)
331. Not to turn one's attention to idolatry (Lev. 19:4)
332. Not to adopt the institutions of idolaters nor their customs (Lev. 18:3; Lev. 20:23)

333. Not to pass a child through the fire to Molech (Lev. 18:21)
334. Not to suffer any one practicing witchcraft to live (Ex. 22:17)
335. Not to practice observing times or seasons -i.e. astrology (Lev. 19:26)
336. Not to practice superstitions/witchcraft (doing things based on signs and potions; using charms and incantations) (Lev. 19:26)
337. Not to consult familiar spirits or ghosts (Lev. 19:31)
338. Not to consult wizards (Lev. 19:31)
339. Not to practice specific magic by using stones herbs or objects. (Deut. 18:10)
340. Not to practice magical practices in general.(Deut. 18:10)
341. Not to practice the art of casting spells over snakes and scorpions (Deut. 18:11)
342. Not to enquire of a familiar spirit or ghost (Deut. 18:11)
343. Not to seek the dead (Deut. 18:11)
344. Not to enquire of a wizard) (Deut. 18:11)
345. Not to remove the entire beard, like the idolaters (Lev. 19:27)
346. Not to round the corners of the head, as the idolatrous priests do (Lev. 19:27)
347. Not to cut oneself or make incisions in one's flesh in grief, like the idolaters (Lev. 19:28; Deut. 14:1)
348. Not to tattoo the body like the idolaters (Lev.

19:28)
349. Not to make a bald spot for the dead (Deut. 14:1)
350. Not to plant a tree for worship (Deut. 16:21)
351. Not to set up a pillar (for worship) (Deut. 16:22)
352. Not to show favor to idolaters (Deut. 7:2)
353. Not to make a covenant with the seven (Canaanite, idolatrous) nations (Ex. 23:32; Deut. 7:2)
354. Not to settle idolaters in our land (Ex. 23:33)
355. To slay the inhabitants of a city that has become idolatrous and burn that city (Deut. 13:16-17)
356. Not to rebuild a city that has been led astray to idolatry (Deut.13:17)
357. Not to make use of the property of city that has been so led astray (Deut. 13:18)

Agriculture and Animal Husbandry

358. Not to cross-breed cattle of different species (Lev. 19:19)
359. Not to sow different kinds of seed together in one field (Lev.19:19)
360. Not to eat the fruit of a tree for three years from the time it was planted (Lev. 19:23)
361. That the fruit of fruit-bearing trees in the fourth year of their planting shall be sacred like the second tithe and eaten in Jerusalem (Lev. 19:24)
362. Not to sow grain or herbs in a vineyard (Deut. 22:9)
363. Not to eat the produce of diverse seeds sown in a

vineyard (Deut. 22:9)
364. Not to work with beasts of different species, yoked together (Deut. 22:10)

Clothing

365. That a man shall not wear women's clothing (Deut. 22:5)
366. That a woman should not wear men's clothing (Deut. 22:5)
367. Not to wear garments made of wool and linen mixed together (Deut. 22:11)

The Firstborn

368. To redeem the firstborn human male (Ex. 13:13; Ex. 34:20; Num. 18:15)
369. To redeem the firstling of an ass (Ex. 13:13; Ex. 34:20)
370. To break the neck of the firstling of an ass if it is not redeemed (Ex. 13:13; Ex. 34:20)
371. Not to redeem the firstling of a clean beast (Num. 18:17)

High Priest, Priests and Levites

372. That the Priest shall put on priestly vestments for the service (Ex. 28:2)
373. Not to tear the High Priest's robe (Ex. 28:32)
374. That the Priest shall not enter the Sanctuary at all times (i.e., at times when he is not performing service) (Lev. 16:2)
375. That the ordinary Priest shall not defile himself by

contact with any dead, other than immediate relatives (Lev. 21:1-3)

376. That the sons of Aaron defile themselves for their deceased relatives (by attending their burial), and mourn for them like other Israelites, who are commanded to mourn for their relatives (Lev. 21:3)

377. That a Priest who had an immersion during the day (to cleanse him from his uncleanness) shall not serve in the Sanctuary until after sunset (Lev. 21:6)

378. That a Priest shall not marry a divorced woman (Lev. 21:7)

379. That a Priest shall not marry a harlot (Lev. 21:7)

380. That a Priest shall not marry a profaned woman (Lev. 21:7)

381. To show honor to a Priest, and to give him precedence in all things that are holy (Lev. 21:8)

382. That a High Priest shall not defile himself with any dead, even if they are relatives (Lev. 21:11)

383. That a High Priest shall not go (under the same roof) with a dead body (Lev. 21:11)

384. That the High Priest shall marry a virgin (Lev. 21:13)

385. That the High Priest shall not marry a widow (Lev. 21:14)

386. That the High Priest shall not cohabit with a widow, even without marriage, because he profanes her (Lev. 21:15)

387. That a person with a physical blemish shall not serve (in the Sanctuary) (Lev. 21:17)

388. That a Priest with a temporary blemish shall not serve there (Lev. 21:21)
389. That a person with a physical blemish shall not enter the Sanctuary further than the altar (Lev. 21:23)
390. That a Priest who is unclean shall not serve (in the Sanctuary) (Lev. 22:2-3)
391. To send the unclean out of the Camp, that is, out of the Sanctuary (Num. 5:2)
392. That a Priest who is unclean shall not enter the courtyard (Num. 5:2-3) This refers to the Camp of the Sanctuary
393. That the sons or descendants of Aaron shall bless Israel (Num. 6:23)
394. To set apart a portion of the dough for the Priest (Num.15:20)
395. That the Levites shall not occupy themselves with the service that belongs to the sons of Aaron, nor the sons of Aaron with that belonging to the Levites (Num. 18:3)
396. That one not a descendant of Aaron in the male line shall not serve (in the Sanctuary) (Num. 18:4-7)
397. That the Levite shall serve in the Sanctuary (Num. 18:23)
398. To give the Levites cities to dwell in, these to serve also as cities of refuge (Num. 35:2)
399. That none of the tribe of Levi shall take any portion of territory in the land (of Israel) (Deut. 18:1)
400. That none of the tribe of Levi shall take any share

of the spoil (at the conquest of the Promised Land) (Deut. 18:1)

401. That the sons of aaron shall serve in the Sanctuary in divisions, but on festivals, they all serve together (Deut. 18:6-8)

Tithes, Taxes and T'rumah [Hebrew Offerings]

402. That an uncircumcised person shall not eat of the t'rumah (heave offering), and the same applies to other holy things. This rule is inferred from the law of the Paschal offering, by similarity of phrase (Ex. 12:44-45 and Lev. 22:10) but it is not explicitly set forth in the Torah. Traditionally, it has been learnt that the rule that the uncircumcised must not eat holy things is an essential principle of the Torah and not an enactment of the Scribes

403. Not to alter the order of separating the t'rumah and the tithes; the separation be in the order first-fruits at the beginning, then the t'rumah, then the first tithe, and last the second tithe (Ex.22:28)

404. To give half a shekel every year (to the Sanctuary for provision of the public sacrifices) (Ex. 30:13)

405. That a preist [kohein] who is unclean shall not eat of the t'rumah (Lev.22:3-4)

406. That a person who is not a kohein or the wife or unmarried daughter of a kohein shall not eat of the t'rumah (Lev. 22:10)

407. That a sojourner with a kohein or his hired servant

shall not eat of the t'rumah (Lev. 22:10)

408. Not to eat unholy things [Heb. tevel] (something from which the t'rumah and tithe have not yet been separated) (Lev. 22:15)

409. To set apart the tithe of the produce (one tenth of the produce after taking out t'rumah) for the Levites (Lev. 27:30; Num.18:24)

410. To tithe cattle (Lev. 27:32)

411. Not to sell the tithe of the heard (Lev. 27:32-33)

412. That the Levites shall set apart a tenth of the tithes, which they had received from the Israelites, and give it to the Priest [Heb.kohanim] (called the t'rumah of the tithe) (Num. 18:26)

413. Not to eat the second tithe of cereals outside Jerusalem (Deut.12:17)

414. Not to consume the second tithe of the vintage outside of Jerusalem (Deut. 12:17)

415. Not to consume the second tithe of the oil outside of Jerusalem (Deut. 12:17)

416. Not to forsake the Levites (Deut. 12:19); but their gifts (dues) should be given to them, so that they might rejoice therewith on each and every festival

417. To set apart the second tithe in the first, second, fourth and fifth years of the sabbatical cycle to be eaten by its owner in Jerusalem (Deut. 14:22)

418. To set apart the second tithe in the third and sixth year of the sabbatical cycle for the poor (Deut. 14:28-29)

419. To give the kohein [i.e. Priest] the due portions of

the carcass of cattle (Deut. 18:3)
420. To give the first of the fleece to the priest (Deut. 18:4)
421. To set apart a small portion of the grain, wine and oil for the Priest [Heb. kohein] [Heb. t'rumah g'dolah i.e.(the great heave-offering) (Deut.18:4)
422. Not to expend the proceeds of the second tithe on anything but food and drink (Deut. 26:14)
423. Not to eat the Second Tithe, even in Jerusalem, in a state of uncleanness, until the tithe had been redeemed (Deut. 26:14)
424. Not to eat the Second Tithe, when mourning (Deut. 26:14)
425. To make the declaration, when bringing the second tithe to the Sanctuary (Deut. 26:13)

The Temple, the Sanctuary and Sacred Objects

426. Not to build an altar of hewn stone (Ex. 20:22)
427. Not to mount the altar by steps (Ex. 20:23)
428. To build the Sanctuary (Ex. 25:8)
429. Not to remove the staves from the Ark (Ex. 25:15)
430. To set the showbread and the frankincense before the Lord every Sabbath (Ex. 25:30)
431. To kindle lights in the Sanctuary (Ex. 27:21)
432. That the breastplate shall not be loosened from the ephod (Ex.28:28)
433. To offer up incense twice daily (Ex. 30:7)
434. Not to offer strange incense nor any sacrifice

upon the golden altar (Ex. 30:9)
435. That the Priest shall wash his hands and feet at the time of service (Ex. 30:19)
436. To prepare the oil of anointment and anoint high priests and kings with it (Ex. 30:31)
437. Not to compound oil for lay use after the formula of the anointing oil (Ex. 30:32-33)
438. Not to anoint a stranger with the anointing oil (Ex. 30:32)
439. Not to compound anything after the formula of the incense (Ex.30:37)
440. That he who, in error, makes unlawful use of sacred things, shall make restitution of the value of his trespass and add a fifth (Lev. 5:16)
441. To remove the ashes from the altar (Lev. 6:3)
442. To keep fire always burning on the altar of the burnt-offering (Lev. 6:6)
443. Not to extinguish the fire on the altar (Lev. 6:6)
444. That a kohein shall not enter the Sanctuary with disheveled hair (Lev. 10:6)
445. That a kohein shall not enter the Sanctuary with torn garments (Lev. 10:6)
446. That the kohein shall not leave the Courtyard of the Sanctuary, during service (Lev. 10:7)
447. That an intoxicated person shall not enter the Sanctuary nor give decisions in matters of the Law (Lev. 10:9-11)
448. To revere the Sanctuary (Lev. 19:30) (today, this applies to synagogues)

449. That when the Ark is carried, it should be carried on the shoulder (Num. 7:9)
450. To observe the second Passover (Num. 9:11)
451. To eat the flesh of the Paschal lamb on it, with unleavened bread and bitter herbs (Num. 9:11)
452. Not to leave any flesh of the Paschal lamb brought on the second Passover until the morning (Num. 9:12)
453. Not to break a bone of the Paschal lamb brought on the second Passover (Num. 9:12)
454. To sound the trumpets at the offering of sacrifices and in times of trouble (Num. 10:9-10)
455. To watch over the edifice continually (Num. 18:2)
456. Not to allow the Sanctuary to remain unwatched (Num. 18:5)
457. That an offering shall be brought by one who has in error committed a trespass against sacred things, or robbed, or lain carnally with a bond-maid betrothed to a man, or denied what was deposited with him and swore falsely to support his denial.
This is called a guilt-offering for a known trespass (Lev. 5:15-19)
458. Not to destroy anything of the Sanctuary, of synagogues, or of houses of study, nor erase the holy names (of God); nor may sacred scriptures be destroyed (Deut. 12:2-4)

Sacrifices and Offerings

459. To sanctify the firstling of clean cattle and offer

it up (Ex. 13:2;Deut. 15:19)
460. To slay the Paschal lamb (Ex. 12:6)
461. To eat the flesh of the Paschal sacrifice on the night of the fifteenth of Nissan (Ex. 12:8)
462. Not to eat the flesh of the Paschal lamb raw or sodden (Ex.12:9)
463. Not to leave any portion of the flesh of the Paschal sacrifice until the morning unconsumed (Ex. 12:10)
464. Not to give the flesh of the Paschal lamb to an Israelite who had become an apostate (Ex. 12:43)
465. Not to give flesh of the Paschal lamb to a stranger who lives among you to eat (Ex. 12:45)
466. Not to take any of the flesh of the Paschal lamb from the company's place of assembly (Ex. 12:46)
467. Not to break a bone of the Paschal lamb (Ex. 12:46)
468. That the uncircumcised shall not eat of the flesh of the Paschal lamb (Ex. 12:48)
469. Not to slaughter the Paschal lamb while there is leaven in the home (Ex. 23:18; Ex. 24:25)
470. Not to leave the part of the Paschal lamb that should be burnt on the altar until the morning, when it will no longer be fit to be burnt (Ex. 23:18; Ex. 24:25)
471. Not to go up to the Sanctuary for the festival without bringing an offering (Ex. 23:15)
472. To bring the first fruits to the Sanctuary (Ex. 23:19)
473. That the flesh of a sin-offering and guilt-offering

shall be eaten (Ex. 29:33)
474. That one not of the seed of Aaron, shall not eat the flesh of the holy sacrifices (Ex. 29:33)
475. To observe the procedure of the burnt-offering (Lev. 1:3)
476. To observe the procedure of the meal-offering (Lev. 2:1)
477. Not to offer up leaven or honey (Lev. 2:11)
478. That every sacrifice be salted (Lev. 2:13)
479. Not to offer up any offering unsalted (Lev. 2:13)
480. That the Court of Judgment shall offer up a sacrifice if they have erred in a judicial pronouncement (Lev. 4:13)
481. That an individual shall bring a sin-offering if he has sinned in error by committing a transgression (Lev. 4:27-28)
482. To offer a sacrifice of varying value in accordance with one's means (Lev. 5:7)
483. Not to sever completely the head of a fowl brought as a sin-offering (Lev. 5:8)
484. Not to put olive oil in a sin-offering made of flour (Lev. 5:11)
485. Not to put frankincense on a sin-offering made of flour (Lev. 5:11)
486. That an individual shall bring an offering if he is in doubt as to whether he has committed a sin for which one has to bring a sin-offering. (Lev. 5:17-19)
487. That the remainder of the meal offerings shall be eaten (Lev. 6:9)

488. Not to allow the remainder of the meal offerings to become leavened (Lev. 6:10)
489. That the High Priest [Heb. Kohein] shall offer a meal offering daily (Lev. 6:13)
490. Not to eat of the meal offering brought by Aaron and his sons (Lev. 6:16)
491. To observe the procedure of the sin-offering (Lev. 6:18)
492. Not to eat of the flesh of sin offerings, the blood of which is brought within the Sanctuary and sprinkled towards the Veil (Lev. 6:23)
493. To observe the procedure of the guilt-offering (Lev. 7:1)
494. To observe the procedure of the peace-offering (Lev. 7:11)
495. To burn meat of the holy sacrifice that has remained over (Lev. 7:17)
496. Not to eat of sacrifices that are eaten beyond the appointed time for eating them (Lev. 7:18)
497. Not to eat of holy things that have become unclean (Lev. 7:19)
498. To burn meat of the holy sacrifice that has become unclean (Lev. 7:19)
499. That a person who is unclean shall not eat of things that are holy (Lev. 7:20)
500. A Priest's daughter who profaned herself shall not eat of the holy things, neither of the heave offering nor of the breast, nor of the shoulder of peace offerings (Lev. 10:14, Lev. 22:12)

501. That a woman after childbirth shall bring an offering when she is clean (Lev. 12:6)
502. That the leper shall bring a sacrifice after he is cleansed (Lev.14:10)
503. That a man having an issue shall bring a sacrifice after he is cleansed of his issue (Lev. 15:13-15)
504. That a woman having an issue shall bring a sacrifice after she is cleansed of her issue (Lev. 15:28-30)
505. To observe, on Yom Kippur, the service appointed for that day, regarding the sacrifice, confessions, sending away of the scapegoat, etc. (Lev. 16:3-34)
506. Not to slaughter beasts set apart for sacrifices outside (the Sanctuary) (Lev. 17:3-4)
507. Not to eat flesh of a sacrifice that has been left over (beyond the time appointed for its consumption) (Lev. 19:8)
508. Not to sanctify blemished cattle for sacrifice on the altar (Lev.22:20) This text prohibits such beasts being set apart for sacrifice on the altar
509. That every animal offered up shall be without blemish (Lev.22:21)
510. Not to inflict a blemish on cattle set apart for sacrifice (Lev.22:21)
511. Not to slaughter blemished cattle as sacrifices (Lev. 22:22)
512. Not to burn the limbs of blemished cattle upon the altar (Lev.22:22)
513. Not to sprinkle the blood of blemished cattle upon

the altar (Lev. 22:24)
514. Not to offer up a blemished beast that comes from non-Israelites (Lev. 22:25)
515. That sacrifices of cattle can only take place when they are at least eight days old (Lev. 22:27)
516. Not to leave any flesh of the thanksgiving offering until the morning (Lev. 22:30)
517. To offer up the meal-offering of the Omer on the morrow after the first day of Passover, together with one lamb (Lev. 23:10)
518. Not to eat bread made of new grain before the Omer of barley has been offered up on the second day of Passover (Lev. 23:14)
519. Not to eat roasted grain of the new produce before that time (Lev. 23:14)
520. Not to eat fresh ears of the new grain before that time (Lev. 23:14)
521. To bring on wave loaves of bread together with the sacrifices which are then offered up in connection with the loaves [Pentecost feast] (Lev. 23:17-20)
522. To offer up an additional sacrifice on Passover (Lev. 23:36)
523. That one who vows to the Lord the monetary value of a person shall pay the amount appointed in the Scriptural portion (Lev. 27:2-8)
524. If a beast is exchanged for one that had been set apart as an offering, both become sacred (Lev. 27:10)
525. Not to exchange a beast set aside for sacrifice (Lev. 27:10)

526. That one who vows to the Lord the monetary value of an unclean beast shall pay its value (Lev. 27:11-13)
527. That one who vows the value of a his house shall pay according to the appraisal of the Priest (Lev. 27:11-13)
528. That one who sanctifies to the Lord a portion of his field shall pay according to the estimation appointed in the Scriptural portion (Lev. 27:16-24)
529. Not to transfer a beast set apart for sacrifice from one class of sacrifices to another (Lev. 27:26)
530. To decide in regard to dedicated property as to which is sacred to the Lord and which belongs to the Priest (Lev. 27:28)
531. Not to sell a field devoted to the Lord (Lev. 27:28)
532. Not to redeem a field devoted to the Lord (Lev. 27:28)
533. To make confession before the Lord of any sin that one has committed, when bringing a sacrifice and at other times (Num. 5:6-7)
534. Not to put olive oil in the meal-offering of a woman suspected of adultery
(Num. 5:15)
535. Not to put frankincense on it (Num. 5:15)
536. To offer up the regular sacrifices daily (two lambs as burnt offerings) (Num. 28:3)
537. To offer up an additional sacrifice every Sabbath (two lambs) (Num. 28:9)
538. To offer up an additional sacrifice every New Moon (Num. 28:11)

539. To bring an additional offering on the day of the first fruits [Pentecost] (Num. 28:26-27)
540. To offer up an additional sacrifice on [Feast of Trumpets] or Rosh Hashanah (Num. 29:1-6)
541. To offer up an additional sacrifice on the day of Atonement or Yom Kippur (Num. 29:7-8)
542. To offer up an additional sacrifice on Feast of Tabernacles [Heb. Sukkot] (Num. 29:12-34)
543. To offer up an additional offering on the eighth day after the feast of Tabernacles called (Heb. Shemini Atzeret), which is a festival by itself (Num. 29:35-38) This eighth day is an anticipation of the New Testament Sabbath which would be instituted on the first day of the week, which is also the eighth day.
544. To bring all offerings, whether obligatory or freewill, on the first festival after these were incurred (Deut. 12:5-6)
545. Not to offer up sacrifices outside (the Sanctuary) (Deut. 12:13)
546. To offer all sacrifices in the Sanctuary (Deut. 12:14)
547. To redeem cattle set apart for sacrifices that contracted disqualifying blemishes, after which they may be eaten by anyone. (Deut. 12:15)
548. Not to eat of the unblemished firstling outside Jerusalem (Deut.12:17)
549. Not to eat the flesh of the burnt-offering (Deut. 12:17). This is a Prohibition applying to every

trespasser, not to enjoy any of the holy things.
550. That the sons of Aaron [i.e. his descendents] shall not eat the flesh of the sin-offering or guilt-offering outside the Courtyard (of the Sanctuary) (Deut.12:17)
551. Not to eat of the flesh of the sacrifices that are holy in a minor degree, before the blood has been sprinkled (on the altar), (Deut. 12:17)
552. That the Priest shall not eat the first-fruits before they are set down in the Courtyard (of the Sanctuary) (Deut. 12:17)
553. To take trouble to bring sacrifices to the Sanctuary from places outside the land of Israel (Deut. 12:26)
554. Not to eat the flesh of beasts set apart as sacrifices, that have been rendered unfit to be offered up by deliberately inflicted blemish (Deut. 14:3)
555. Not to do work with cattle set apart for sacrifice (Deut. 15:19)
556. Not to shear beasts set apart for sacrifice (Deut. 15:19)
557. Not to leave any portion of the festival offering brought on the fourteenth of Nissan unto the third day (Deut. 16:4)
558. Not to offer up a beast that has a temporary blemish (Deut.17:1)
559. Not to bring sacrifices out of the hire of a harlot or price of a dog (apparently a euphemism for sodomy) (Deut. 23:19)

560. To read the portion prescribed on bringing the first fruits (Deut.26:5-10)

Ritual Purity and Impurity

561. That eight species of creeping things defile by contact (Lev.11:29-30)
562. That foods become defiled by contact with unclean things (Lev.11:34)
563. That anyone who touches the carcass of a beast that died of itself shall be unclean (Lev. 11:39)
564. That a lying-in woman is unclean like a menstruating woman (in terms of uncleanness) (Lev. 12:2-5)
565. That a leper is unclean and defiles (Lev. 13:2-46)
566. That the leper shall be universally recognized as such by the prescribed marks So too, all other unclean persons should declare themselves as such (Lev. 13:45)
567. That a leprous garment is unclean and defiles (Lev. 13:47-49)
568. That a leprous house defiles (Lev. 14:34-46)
569. That a man, having a running issue, defiles (Lev. 15:1-15)
570. That the seed of copulation defiles (Lev. 15:16)
571. That purification from all kinds of defilement shall be effected by ceremonial washing (Lev. 15:16)
572. That a menstruating woman is unclean and defiles others (Lev.15:19-24)
573. That a woman, having a running issue, defiles (Lev.

15:25-27)
574. To carry out the ordinance of the Red Heifer so that its ashes will always be available (Num. 19:9)
575. That a corpse defiles (Num. 19:11-16)
576. That the waters of separation defile one who is clean, and cleanse the unclean from pollution by a dead body (Num.19:19-22)

Lepers and Leprosy

577. Not to drove off the hair of the scalp (Lev. 13:33)
578. That the procedure of cleansing leprosy, whether of a man or of a house, takes place with cedar-wood, hyssop, scarlet thread, two birds, and running water (Lev. 14:1-7)
579. That the leper shall shave all his hair (Lev. 14:9)
580. Not to pluck out the marks of leprosy (Deut. 24:8)

The King

581. Not to curse a ruler, that is, the King in the land of Israel (Ex. 22:27)
582. To appoint a king (Deut. 17:15)
583. Not to appoint as ruler over Israel, one who comes from non-Israelites (Deut. 17:15)
584. That the King shall not acquire an excessive number of horses (Deut. 17:16)
585. That the King shall not take an excessive number of wives (Deut. 17:17)

586. That he shall not accumulate an excessive quantity of gold and silver (Deut. 17:17)
587. That the King shall write a scroll of the Torah for himself, in addition to the one that every person should write, so that he writes two scrolls (Deut. 17:18)

Nazarites

588. That a Nazarite shall not drink wine, or anything mixed with wine which tastes like wine; and even if the wine or the mixture has turned sour, it is prohibited to him (Num. 6:3)
589. That he shall not eat fresh grapes (Num. 6:3)
590. That he shall not eat dried grapes (raisins) (Num. 6:3)
591. That he shall not eat the kernels of the grapes (Num. 6:4)
592. That he shall not eat of the skins of the grapes (Num. 6:4)
593. That the Nazarite shall permit his hair to grow (Num. 6:5)
594. That the Nazarite shall not cut his hair (Num. 6:5)
595. That he shall not enter any covered structure where there is a dead body (Num. 6:6)
596. That a Nazarite shall not defile himself for any dead person (by being in the presence of the corpse) (Num. 6:7)
597. That the Nazarite shall shave his hair when he brings his offerings at the completion of the period of his Naziriteship, or within that period if he has become defiled (Num. 6:9)

Wars

598. That those engaged in warfare shall not fear their enemies nor be panic-stricken by them during battle (Deut. 3:22, 7:21, 20:3)
599. To anoint a special Priest (to speak to the soldiers) in a war (Deut. 20:2) This is today's equivalent to a military chaplain.
600. In a permissive war (as distinguished from obligatory ones), to observe the procedure prescribed in the Torah (Deut. 20:10)
601. Not to keep alive any individual of the seven Canaanite nations (Deut. 20:16)
602. **To exterminate the seven Canaanite nations from the land of Israel (Deut. 20:17)** (*If Israel had done this the Jew would not be a problem today!*)
603. Not to destroy fruit trees (wantonly or in warfare) (Deut. 20:19-20)
604. To deal with a beautiful woman taken captive in war in the manner prescribed in the Torah (Deut. 21:10-14)
605. Not to sell a beautiful woman, (taken captive in war) (Deut. 21:14)
606. Not to degrade a beautiful woman (taken captive in war) to the condition of a bondwoman (Deut. 21:14)
607. Not to offer peace to the Ammonites and the Moabites before waging war on them, as should be done to other nations (Deut. 23:7)
608. That anyone who is unclean shall not enter the

Camp of the Levites (Deut. 23:11)
609. To have a place outside the camp for sanitary purposes (Deut. 23:13)
610. To keep that place sanitary (Deut. 23:14-15)
611. Always to remember what Amalek did (Deut. 25:17)
612. That the evil done to us by Amalek shall not be forgotten (Deut. 25:19)
613. To destroy the seed of Amalek (Deut. 25:19)

THE TALMUD
A STUDY IN TOLERANCE AND UNDERSTANDING
[WHAT THE "JEWS" REALLY BELIEVE]

Sanhedrin 78a: Allows a Jew to have intercourse with a dead body!

Kallah, 1b, (18b): "Jesus was illegitimate and conceived during menstruation."

Scabbat XIV: "Jesus is referred to as the son of a Roman soldier and a Jewish Prostitute."

Sanhedrin, 103a: "This passage suggests that Christ corrupted His morals and dishonored Himself."

Sanhedrin, 107b: "This passage states that Christ seduced and destroyed Israel."

Hilkoth Melakhim: Suggests that Christians sin by worshipping Jesus Christ.

Hilkoth Maakhaloth: "Christians are idolaters."

Orachrach Chaiim (20, 2): "Christians disguise themselves as Jews in order to kill them."

Abhodah Zorah (15b): Suggests that Christians have sexual relations with animals.

Babha Kama (113b): "The name of God not profaned, if a Jew lies to a Christian."

Tract Mechilla: "Almighty God studies the Talmud standing, because He has such respect for that book."

Sanhedrin (59a) & Abohodan Zarah 8-6: "Every goy (non-Jew) who studies the Talmud and every Jew who helps him in it, ought to die."

Szaaloth-Utszabot, The Book of Jore Dia 17: "A Jew should and must make a false oath when the goyim asks if our books contain anything against them."

Schulchan Aruch, Choszen Hamiszpat 348: "A Jew may rob a goy (non-Jew) that is, he may cheat him in a bill, if unlikely to be perceived by him."

Simeon Haddaesen fol. 56-D: "When the Messiah comes every Jew will have 2800 slaves."

Kelhubath (11a-11b): "When a grown-up man has had intercourse with a little girl...It means this: WHEN A GROWN UP MAN HAS INTERCOURSE WITH A LITTLE GIRL IT IS NOTHING, FOR WHEN THE GIRL IS LESS THAN THIS THREE YEARS OLD IT IS AS IF ONE PUTS THE FINGER INTO THE EYE tears come to the eye again and again, so does virginity come back to the Little Girl Three Years Old." (Why are you surprised Jews are Satan's Children could you expect less?)

Midrasch Talpioth 225-L: "Jehovah created the non-Jew in human form so that the Jew would not have to be served by beasts. The non-Jew is consequently an animal in human form, and condemned to serve the Jew day and night."

Nadarine, 20, B; Schulchan Aruch, Choszen Hamiszpat 348: "A Jew may do to a non-Jewess what he can do. He may treat her as he treats a piece of meat."

Josiah 60, 6, Rabbi Abarbanel to Daniel 7, 13: "As soon as the King Messiah will declare himself, and He will destroy Rome and make a wilderness of it. Thorns and weeds will grow in the Pope's palace. Then He will start a merciless war on non-Jews and will overpower them. He will slay them in masses, kill their kings and lay waste the whole Roman land. He will say to the Jews: 'I am the King Messiah for whom you have been waiting. Take the silver and gold from the goyim."

Chaggigah, (15b): "A Jew is considered to be good in the eyes of God, in spite of any sins he may commit."

Schulchan Aruch, Choszen Hamiszpat 348: "All property of other nations belongs to the Jewish nation, which, consequently, is entitled to seize upon it without any scruples (This is what the Jews use for justification to steal the land of the Palestinians). An orthodox Jew is not bound to observe principles of morality towards people of other tribes. He may act contrary to mortality, if profitable to himself or to Jews in general."

Tosefta, Abhodah Zarah VIII, 5: "How to interpret the word 'robbery.' A goy (non-Jew) is forbidden to steal, rob, or take women slaves, etc., from a goy or from a Jew. But a Jew is Not forbidden to do all this to a goy."

Schulchan Aruch Edit, I, 136: "All vows, oaths, promises, engagements, and swearing, which, beginning this very day of reconciliation till the next...we intend to vow, promise, swear, and bind ourselves to fulfill, we repent of beforehand; let them be illegalized, acquitted, annihilated, abolished, valueless, unimportant. Our vows shall be no vows, and our oaths no oaths at all."

Schulchan, Aruch Orach Chaim 539: "At the time of the Cholhamoed the transaction of any kind of business is forbidden. But it is permitted to cheat a goy (non-Jew), because cheating of goy at any time pleases the Lord."

Schulchan, Aruch Choszen Hamiszpat 388: "It is permitted to 'Kill a Jewish Denunciator everywhere. It is permitted to kill him even before he denounces."

Livore David 37: "If a Jew be called upon to explain any part of the rabbinic books, he ought to give only a false explanation. Who ever will violate this order shall be put to death."

Abhodah Zarah 26b Tosephoth: "A Jew who kills a Christian commits no sin, but offers an acceptable sacrifice to God."

Tosefta, Erubin VIII, 1: "On the house of the goy (non-Jew) one looks as on the fold of cattle."

Sanhedrin 67a: Jesus referred to as the son of Pandira, a soldier. Mother a prostitute.

Kallah 1b. (18b): Christ illegitimate and conceived during menstruation. Mother a Prostitute.

Sanhedrin 67a: Jesus was hanged on the eve of Passover.

Toldath Jeschu: The Birth of Christ related in most shameful expressions.

Abhodah Zarah II: Referred to as the son of Pandira, a Roman soldier, a
Prostitute Mother.

Schabbath XIV: Again referred to as the son of Pandira the Roman soldier.

Sanhedrin 43a: On the eve of Passover they hanged Jesus.

Schabbath 104b: Christ called a fool and no one pays attention to fools.

Toldoth Jeschu: Says Judas and Jesus engaged in a quarrel with human
excrement.

Sanhedrin 103a: Suggested corrupts his morals and dishonors self.

Sanhedrin 107b: Seduced, corrupted and destroyed Israel.

Zohar III (282): Died like a beast and buried in animal's dung heap.

Hilkoth Melakhim: Attempt to prove Christians err in worship of

Jesus.

Abhodah Zarah 21a: Reference to worship of Jesus in homes unwanted.

Orach Chaiim 113: Avoid appearance of paying respect to Jesus.

Iore Dea 150, 2: Do not appear to pay respect to Jesus by accident.

Abhodah Zarah (6a): False teaching to worship on the first day of Sabbath.

Kerithuth (6b p. 78): Jews called men, Christians are not called men.

Makkoth (7b): Innocent of murder if intent was to kill Christian.

Sohar (II 64b): Christian birth rate must be diminished materially. {Now you know why the Jews are always pushing for abortions}.

Schabbath (116a) Tos: Gospels called volumes of iniquity, heretical books.

Schabbath (116a): Talmudists agree that the books of Christians are to be burned.

Chullin (91b): Jews possess dignity even an angel cannot share.

Hilkoth Akum (V. 12): Quote Scriptures Forbid mentioning the Christian
God.

Choschen Ham (226 1): Jew may keep lost property of Christian found by
Jew.

Babba Kama (113b): It is permitted to deceive Christians; Jew may lie and perjure to condemn a Christian; Name of God not profaned when lying to Christians.

Kallah (1b p. 18): Jew may perjure himself with a clear conscience.

Schabbouth Hag. (d): Jews may swear falsely with subterfuge wording.

Zohar (1 160a): Jews must always try to deceive Christians.

Choschen Ham (425 5): Jews are not to prevent the death of a Christian.

Hilkkoth Akum (x,1): Do not save Christians in danger of death, instructed to let die.

Abhodah Zarah (25b)T: Even the best of the Goyim (Christians) should be
killed.

Sepher or Israel (177b): If Jew kills a Christian he commits no sin.

Zohar (11 43a): Extermination of Christians necessary.

Hilkhpth Akum (x,1): Make no agreements and show no mercy to Christians.

Hilkhoth Maakhaloth: Christians are idolaters.

Iore Dea (198, 48): Female Jews contaminated when meeting Christians.

Makkoth (7b): Innocent of murder if intent was to kill a Christian.

Orach Chaiim (225, 10): Christians and animals grouped for comparisons.

Zohar II (64b): Christians likened to cows and asses.

Kethuboth (110b): Psalmist compares Christians to beasts.

Sanhedrin (74b) Tos: Sexual intercourse with Christian same as intercourse with beast.

Iore Dea (337, 1): Replace dead Christians like you would a lost cow or ass.

Chullin (91b): Jews possess dignity even an angel cannot share.

Sanhedrin (58b): To strike a Jew is the same as slapping the face of God.

Zohar (1, 25b): Those Jews who do good to Christians never rise when dead.

Iore Dea (148, 12H): Jews are to hide their hatred for Christians.

Babha Bathra (54b): Christian property belongs to the first Jew claiming it.

Babha Kama (113b): It is permitted for a Jew to deceive Christians.

Iore Dea (157, 2) H: Jew may deceive Christians.

Babha Kama (113a): Jew may lie and perjure himself to condemn a Christian.

Babha Kama (113b): The name of God is not profaned when Jew lies to Christians.

Kallah (1b, p. 18): Jew may perjure himself when lying about Christians.

Schabbouth Hag (6d): Jews may swear falsely by the use of subterfuge wording.

Zohar (1, 160a): Jews must always try to deceive Christians.

Choschen Ham (425, 5): Do not prevent a Christians death.

Iore Dea (158, 1): Christians who are Not Jews' enemies Must also die.

Sanhedrin (59a): Christians who study the Jews' "Laws" {Talmud} to be put to death.

Hilkhoth Akum (X,2): Baptized Jews are to be put to death.

Zohar (1, 25a): Christians are to be destroyed when no danger of discovery.

Sepher or Israel (177b): If a Jew kills a Christian he commits no sin. He has done God a service.

Ialkut Simoni (245c): A Jew shedding the blood of a Christian is offering a Sacrifice to God.

Zohar (II, 43a): Extermination of Christians is a necessary sacrifice to God.

Zohar (L, 38b, 39a): A Jew to receive a high place in heaven if he kills a Christian.

Hilkhoth Akum (X,1): Jews are to show no mercy to a Christian.

What's Happened to AmeriKa?

Sometimes I sits and thinks and sometimes I just sits. So goes the old saying from a black man sitting on his porch in the south somewhere.

I have the advantage of looking back more than 60 years with some alacrity and have the time to consume copious numbers of books. I have videos of old WWI, WWII, Korea and Vietnam 'war" movies that were made from original stock and commentaries that challenge the modern revisionists about the how and why of those travesties.

There has been much written and commented on TV about the maladies facing AmeriKa and all skirt the real issues, intentionally or through ignorance. It has been disconcerting to me and the propaganda machine relentlessly pours out the party line defying common sense and logic. After cruising the magazines, newspapers, TV news, the sitcoms, movies and the Internet it is very clear to me that most of AmeriKa under the age of 40 has been thoroughly indoctrinated to the socialist agenda.

And what is this socialist agenda? Sounds rather benign, doesn't it? Socialist. Social. A nice and friendly image. The new mantra is "progressive" and even that sounds just as benign. Progressive; forward thinking, modern. What do these mean and how do these philosophies mold our thinking and thought processes? Not many can articulate the true roots of these "movements". And movements they are. Movements to bring about world wide tyranny control and even the death of many of the people of the world, especially whites, and more to the point, white Christians.

I want to take you back 80 years or so. We were about to enter into a World War on two fronts, Europe and the Pacific. As we began to mobilize and build the materials of war battles began to rage in Europe and Asia. Bombs were dropped and cities burned everywhere and constantly. Our industrial base, chemicals, mining, steel and aluminum were being produced in factories that spewed smoke and heat day and night for years.

Thousands of ships, yes, you read right, thousands of ships and thousands of aircraft were sunk and shot down. All were fully fueled and ultimate leaked into two major oceans and the Mediterranean Sea. Through brute force, productivity, technology and massive manpower we defeated our enemies and those of our allies. Later, when I spent many years in the 50's and 60's transversing the oceans by ship and in the air there was hardly any evidence that a war even happened even in the South Pacific where I spent most of my time. The ocean somehow cleansed itself evidenced by the abundance of sea life I observed everywhere. It was possible to dive in the shallower areas of these oceans and find sunken ships and aircraft from the war. The water is pristine and clear.

There was no evidence of fuel or oil remaining. I find that very interesting in retrospect. The major oil spills that we all have come to know and even the leaks that occur regularly from tankers and other ships are denounced as atrocities and an assault on the environment. These too finally dissipate over time with no evidence that anything untoward ever happened.

Today the "progressive" environmentalists cry out daily about global warming and, particularly, manmade global warming. The hole in the ozone is another bugaboo. They do not have sufficient long term data to make any of these claims so they use computer "modeling" to project the hazards confronting us as a result of this "warming". The ozone layer in the southern hemisphere expands and contracts on a regular basis and is a constant source of speculation about its effect on life on this earth.

As if a computer can predict the future! If a computer could predict the future about anything every "modeler" would be rich because they would know how the markets will react in the future and win nearly always at the casinos. But that they cannot do. What they can do is convince a brain-dead lot of AmeriKans that they are "experts" and their computer models are infallible. What they don't tell you is that data cannot lie but that the data can be rigged. Garbage in, garbage out so the saying goes.

Point? No greater pollution was generated on this planet as was during the 1940's and 1950's and no measurable effects have resulted that we can see today. As a point of fact, the polluters quit polluting in the 1960's and pretty much cleaned up the rivers and environment since then……without any environmentalist intervention! I have a firsthand experience with the smokestacks in Pittsburgh as well as the polluted three rivers during the war and the cleanup afterwards.

There is volcanic activity all over the world. They are spewing fumes and magma day and night. We only hear about major explosions that make for spectacular videos and sometimes cause catastrophic damage. Read this and read it well…..One major volcanic event emits more pollution in one day than all the vehicles and industrial emissions created in America in a year!

Furthermore, there are 25,000 lightning strikes <u>each day</u> starting untold number of forest fires spewing smoke and pollution all over the world. The environmentalist whacko's won't tell you any of this because it will expose them for the idiots that they are. Or are they? No, they have a higher agenda to reduce the West to the likes of a third world country by legislating our businesses to bankruptcy or seek cheap foreign labor in those third worlds so they don't have to concern themselves about pollution. It looks like they have all but succeeded in their agenda.

The unemployment rate is 5% thereabouts according to the government. We know that the government cooks the books every day generating false data with all government agencies and 5th column media in collusion. The real number is over 20% when all eligible workers are considered. Is the government responsible for generating jobs? Absolutely not! The government can only contribute to the financial wellbeing of business or the economy by getting out of it! They not only interfere with the natural order of business and economics, they tax and print money to give to non-producers, including themselves. What we must never forget is that true wealth comes from productivity, never from moving paper or fiat money around. The government prints money which creates inflation. Inflation destroys the future for our children and their children. They will make less and less real income as the cost of goods goes higher.

Am I wrong? Again, I have the advantage of looking at 60 years as an adult to see how our value has eroded. In real dollars from 1960 inflation has risen 1000%! The average worker in AmeriKa earns somewhere around $12.00 per hour. He or she cannot pay rent, purchase a car, go to college and eat anything but fast food on those wages. We have more individuals moving back home or living together than ever before. More and more young men and women are living together without marriage which is another issue to be addressed later. In other words, our productivity, removing those elements from the earth, formulating and combining them into materials, combining those materials into products that are desired and can be utilized by the consumer is what makes nations great. We no longer do any of those things!

Nations became wealthy in the days of old when they colonized to mine gold, silver, agricultural commodities and copper from the "invaded" nation. Others went to war and carried of the booty and slaves to enrich their own treasuries and garner free labor. In all cases, without these spoils they themselves would be overrun because they did not have the resources to field armies and navies to defend and plunder more. Such is the state of nation building without major productivity to create wealth.

When you look at the intellect of the ancients compared to today we are ignoramuses. No, I'm it's not about scientific endeavors. It's about the literal genius evidenced in the writings of Homer, Plato, Tacitus, Aurelius and many others. Even the principles set forth by Archimedes, Pythagoras, Ptolemy and Hippocrates leaves one to wonder how such genius transcended the thinking of the day.

The intellect of the Founders vs. today's offering of weak kneed politicians is a wondrous thing to observe. The Federalist papers between Hamilton, Madison and Jay are a marvel of intellect, reason and thought. Could such a series of arguments be put forth today? I think not. Little known is the fact that Jefferson wrote to King George III many times before he wrote our astounding Declaration of Independence at the age of 27! Franklin was a genius in his own right and went to France with Jefferson to win an ally against England with troops and money. Men among men. Men who can never be emulated in today's world.

Every one of the Founders who signed the Declaration of Independence suffered death or loss of all that they owned by the English. Do we have any in America today who are willing to stand against the tyranny of the government that exists and risk all? Not very likely or they would do it now, this very minute. We are fatted calves, living for celebrity, sports and sex with no sense of propriety any longer. The children that you have who have not been thoroughly indoctrinated by the Marxist/Gramsci educational system will hate you for your complacency and despise you for your cowardess!

Accounting for Catholicism as the nations of Europe were being formed by the Israelites of old nearly all white nations were steeped in the Old Testament and the Law. Christianity became something else after the last apostle died and during the following four generations. Christianity turned apostate in that the tenets put forward by Jesus, the Christ were skewed, distorted and misinterpreted. Even as the revolution and reformation occurred led by Wycliffe, Tyndale, Knox, Calvin, Martin Luther and others, the teachings and truth about white heritage was all but lost until the 1800's. At that time many searched diligently for the truth and found it. The re-birth of the true heritage of whites started to gain hold and many nations of Europe clearly recognized who they were; the descendants of the Patriarchs of the House of Israel, the twelve tribes who migrated from the time of the Egyptian captivity, to the Assyrian Captivity, the Babylonian Captivity and finally, those left in the Southern Kingdom who departed just before Titus destroyed

Jerusalem. Truly a revelation beyond measure!

Even so, the vast majority of whites and particularly white Christians are not aware of their true identity and the awesome responsibility that entails. They, because of the church leadership have fallen away. The leaders have been indoctrinated by their apostate and lying seminaries and Bible Colleges which have been infiltrated by the natural enemy of we whites, the Jew. This fact alone has led the European nations and Amerika down the path to tyranny and ignorance.

The church is entirely apostate today, Catholic and Protestant. They have fallen away and fail to teach the message of real salvation, redemption and are spiritualizing the war that we must fight to regain our nation and identifying the true enemy of Christianity, Lucifer's minions, the Jew. The Jew has two purposes, as few a people as they are in the world.

First, to destroy the white race and Christianity. Their hatred for Jesus the Christ is so embedded in their DNA and gene pool they have no control over this mission.

Second is to rule the world. How is that possible for so few a people? The Jew controls the monetary systems of the world today! They have always been money changers from biblical time and the Rothschild dynasty has had 400 years to accumulate more than 200 trillion dollars in real wealth; gold, silver, real estate, gems and minerals. That is more than any nation or group of nations in the world hold as net worth. He who has the money makes the rules…..so goes the golden rule.

They own the media in most western nations. With that powerful tool they can feed the mindless public drivel and lies each and every day. The pen being mightier than the sword they do not even have to invade their host nations to control thought, elections and laws. Queen Isabella saw it and expelled the Jew. Hitler saw it and forbid the Jew from owning any business or newspaper. The audacious Jew declared war on Germany in 1933 when Hitler was elected Chancellor, long before he began to seize Jew assets. Most, read this carefully, most Jews left Germany before 1933 and before the war. Most left Western Europe for America, Russia and Latin America before the war. So many left there weren't three million left in all of Europe. Hitler must have let them reproduce in droves to have "gassed" 6 million as the lie continues to live. The lie is worth billions to these evil Jews as they extort nations, banks and businesses to this very day. It is too sad that Hitler never accomplished what they say he did. He should have killed them all and

sent his Gestapo to search them out all over the world until every one of them was eliminated. The world would be a far better place!
Here is our dilemma today:

Marxism - Workers unite….Unions

US Bankrupt and owned by the international bankers (Jews) since 1933

Gramsci - Destroy the family, culture and ideals through the children and Marxism will win (Fabianism)

TV Commercials all at the same time (in collusion) to keep from channel changing.

Blacks as a tool to polarize and keep whites off balance and guilt ridden. Always in superior roles on TV thanks to the Jew

Catholicism

Executive Orders and Presidential Directives

Endless unwinnable wars

Open borders

Importing Muslims by the thousands through the back door

Sodomy and fornication rampant and in your face every day and night on TV and from Hollywood

Government out of control spending

Lies and corruption on all levels of government and more flagrant than ever

How can we white Israelites maintain control and dominion over the earth if we give away or sell our creativity and technology to the mixed multitudes!

The reader must be up their ears with the God thing by now. If you have read thus far I have even more to relate with respect to God "things".

As mentioned before God just didn't drop us off to fend for ourselves. He has also protected us in very unique and special ways.

The earth is self-rejuvenating. No matter how much pollution or chemicals or volcanic activity or forest fires or lightning strikes or flood or drought the earth always returns to purity. It is accomplished through the rise of water to the ozone layer and cleansed there. It is accomplished by the powerful sun's rays which are photons, the most powerful energy in the Universe. The lightning strikes releases ozone to purify the moisture and air. There are more than 25,000 lightning strikes a day to do that along with the ozone layer. Oil spills, while highly detrimental to the environment, are nullified over years and are untraceable.

There is the same amount of water on the planet as there was tens of thousands of years ago. It moves about as rivers in the sky, rivers and streams on and under land, ice at both poles and in the winter in the temperate zones and as moisture and rain and in the oceans and seas. No matter how polluted it becomes it rises into the ozone over time and is cleansed to fall back to earth as rain to replenish the supply on the earth! This water must be here so that we live, so that all life may live.

With millions of rocks and boulders and even larger bodies of broken planets, asteroids and comets hurling around outer space at speeds more than a thousand miles a minute we are protected by our atmosphere from all those that would collide with us or us with them. They burn up and are reduced to ash before they strike the earth. Those that do strike the earth are so small as to be a non-event. There have been no meteor strikes in historical times that have caused anything that might resemble catastrophic or even harm!

We see the moon in the heavenly every night and most days. Yet we never consider that since the beginning of recorded history not one person has seen any more of the moon than we see today. Realize that the "face" of the moon is what we see whenever there is a full moon and it is always the same. No one on earth has seen more of the moon than what we see today. It is a miracle in itself that the moon ***makes one rotation only*** each time it revolves around the earth! The moon goes around the earth a little over 27 days and shows us no more than what we see each and every day, each and every week, each and every year, to infinity! The moon has played an important part in the existence of earth and man. The tide changes caused by the moon sustain life in the shoals and shallows which feed higher life forms and so on. Man has used the phases of the moon for planting and sowing, for feast days and fast days, for time keeping and more.

The caterpillar that metamorphoses into a butterfly is amazing and defies all theories of evolution!

The fact that, in any nation, saving the fact of war or epidemic which takes the lives of a large number of the population especially males, the birth of males and females are equal to within 5% is a miracle unto itself!

The fact that the earth is made up of 75% water and we humans are made up of 75% water is amazing.

Flies are a nuisance to all of us, especially when they are buzzing around our homes. Yet, flies are a very necessary part of our ecosystem. They lay eggs that morph into maggots. Maggots thrive on rot and other microorganisms that otherwise could carry disease and other pathogens which would be detrimental to our health, surely an example of the perfection of God's synergism. Contrarily, flies can often carry pathogens as they move from rot to our dinner tables!

Bats abound in certain parts of the country and swarm like locusts to feed on "bugs" that destroy our crops. Little known is that they are nearly as prolific as the honey bee for pollinating our plants. This is especially necessary for blossoms turning into our fruits and vegetables.

We animals have bacteria in our stomach and intestines to ferment our food for assimilation. These "good" bacteria are necessary for our health and wellbeing and another example of God's perfection in our creation.

Every year most trees drop their leaves in the temperate zones of the world. Yet within a year there is no trace of these leaves on the ground. Bacteria transform these leaves into food and soil necessary for feeding the tree nutrition!

The fact that trees all grow straight up even on the sides of mountains or hills gives one to wonder.

The laws of God are consistent and perfect. One that we hear from time to time is "like kind after like kind." Everything procreates after its own species and never mixes species. This is yet another validation of the impossibility of Evolution.

We look at each other outwardly as similar humans and recognize that we each have a personality somewhat unique to each other. What we cannot see is that we are a mass of energy, comprised of atoms each grouping forming all the organs, tissues, bones, eyes, ears, body functions and much, much more and that we emit this energy and receive energy from the ethereal and our surroundings including others. We indeed are spirit beings and our earthly bodies are just a vessel for our energy and composite!

Even if you are a non-believer or skeptic you will be hard presses to proffer a better explanation! **"In the beginning God!"** All designs require a designer and perfect designs can only be from God.

The extensive iteration of a certain peoples noted herein is paramount to a full and comprehensive understanding of the Bible. It is not a mistake that so much of the Bible and who it was written for is so misinterpreted. The teachings in the Seminary's, from the Pulpits and through the media are remiss in identifying the "Chosen" and where they are today. The repetition proffered throughout this book is to reinforce the facts that have been ignored (intentionally?) for centuries. There are those who profit immensely from these falsehoods and are able to control the believing masses with these errors!

If the reader has gone this far and still doubt the veracity of the author and the history of the people of Israel and science obvious by God then nothing will convince him or her. These facts, coupled with those noted earlier, not only prove the existence of God but that He has made everything in perfection, engineered a synergism among all living things, has a plan and purpose for mankind whom He created for His particular pleasure and has everything in motion which I have determined to be

Good "GOD" Vibrations

The End

www.ingramcontent.com/pod-product-compliance
Lightning Source LLC
Chambersburg PA
CBHW070319190526
45169CB00005B/1672